マリタイムカレッジシリーズ

池田 恭子・長山 昌子 編著

KAIBUNDO

はじめに

　本書は，商船系高等専門学校で学んでいるみなさんのために，以下の目標を掲げてつくられました。
　① 海や船に関する仕事を英語で表現できるようになる。
　② 英語を使って世界の人々とコミュニケーションする力を伸ばす。
　③ 英語学習を通じて，仕事や生き方が見えてくる。
　みなさんの多くは，いつか海や船に関する仕事につきたいと思って日々学習や実習に励んでいるかもしれません。
　あるいは，本当にやりたいことは他にあると思っている人もいれば，まだそれが見つからないという人もいると思います。
　そもそも「働く」とか「仕事をする」ということがまだピンとこないという人もいるでしょう。
　しかし，海や船に関する仕事につこうと思っている人はいうまでもなく，将来どのような方面に進むにしろ，現代では英語を使って自由にコミュニケーションする力が必要になっています。
　本書を使って積極的に学ぶことで，英語という武器（あるいは道具）を獲得し，それぞれの夢に向かって飛翔することができるでしょう。
　また，海を舞台とした仕事について学ぶことで，働くことについて，異なる文化背景を持つ人と仕事することについて，より具体的なイメージを持つことができるでしょう。
　本書は，「仕事」「海」「英語」の3つをキーワードとして書かれています。言い換えると，英語を通じて「海」に関わる「仕事」を学ぶことで，目指す目的地がはっきりと見えてくるように工夫されています。
　みなさんが英語という翼を背につけて，大海原を疾走し，夢に向かって空高く舞い上がる姿を確信しています。
　　　　　　Hitch your wagon to a star!　（汝の馬車を星につなげ）
　そんな思いを込めて，本書はつくられました。

謝　辞

　『Navigating English』は，若い学生のみなさんが，海に関わる仕事に興味を持ち，将来的には世界の海に関わりながら働くことに意欲が持てることを目的に執筆・編集されました。

　池田が全体構成，海しごと＆海事知識，ダイアログおよびリーディングの部分を担当し，フォーカスする文法／表現については長山が執筆しました。

　本書の出版は，本当に多くの方々に協力を仰ぎながら，支えていただいたおかげで実現しました。感謝の気持ちでいっぱいです。

　まず，本書に使われているイラストは，大島商船高等専門学校の前畑航平先生がご多忙の中，作成・提供してくださいました。

　本書の作成にあたり，富山高等専門学校商船学科の卒業生，岡島拓哉さん，無関健一朗さん，道渕卓弥さん，長田岳人さん，杉江実宝さんにインタビューを実施しました。彼らは多忙な任務の合間に貴重な休暇を取り，インタビューに協力してくださいました。本書のストーリー構成や取り扱うトピック・事項は彼らの実際の体験や話をもとにつくられたものです。

　また，富山高等専門学校の教職員のみなさまには，海運業界ステークホルダーの方々へのアンケートの実施や多方面への諸連絡など多大なご協力をいただきました。とくに，勝島隆史先生，経田僚昭先生には，航海・機関など専門分野にわたる知識や情報などに関して貴重なアドバイスをいただき，かつ丹念な確認作業をしていただきました。

　最後になりましたが，富山高等専門学校の遠藤真先生には，大学間連携共同教育推進事業「海事分野における高専・産業界連携による人材育成システムの開発」のひとつのプロジェクトとして，本書の企画に関わっていただき，教材開発に向けてさまざまなアイディアをいただきました。心から深く感謝を申し上げます。

　そして，海文堂出版の岩本登志雄氏には本書全般にわたってきめ細かいご配慮をいただきました。

　『Navigating English』は，このように幅広い分野の方々から支えていただいたおかげで出版できました。改めて，みなさまに心よりのお礼を申し上げます。

池田恭子・長山昌子

目　次

はじめに ……………………………………………………3
謝辞 ………………………………………………………4
21世紀型海事人材に求められる能力 ……………………6
本書の構成と利用法 ………………………………………8
登場人物紹介 ……………………………………………10

Story 1	Types of Ship We Operate	12
Story 2	A List of Things to Pack	20
Story 3	Daily Responsibilities and Tasks	29
Story 4	You Must Be …	36
Story 5	Unmanned Engine Room	45
Story 6	Establishing a Good Relationship	54
Story 7	Check the Strainer	62
Story 8	Whenever We Have Newcomers	71
Story 9	This Is My First Watchstanding	80
Story 10	Charted Course Is 055 Degree	91
Story 11	What Are You Going to Serve?	99
Story 12	I Wish I Could Be Home …	111
Story 13	Long Time No See!	120
Story 14	A Logbook Serves as an Important Record	131
Story 15	Sending a Mayday Signal	142
Story 16	We Are Ready for Unmanned Operation	150
Story 17	Welcome Aboard	159
Story 18	I Have Enclosed Our Damage List	168
Story 19	How Do You Read Me?	179
Story 20	Let's Toast!	188

練習の解答例／リーディングの訳 ………………………198

21世紀型海事人材に求められる能力

　私たちが生きる世界は，変化に富んだ，流動的な世界であり，国境を越えて人やものが流れています。とくに海と船は，グローバル化が進む世界経済の大動脈です。また，世界的規模での変化の担い手であると同時に，そのような変化に日々刻々と影響を受ける存在でもあります。そんな世界を現場として働く海事人材には，いったいどんな能力が求められるのでしょうか。

　本書の企画・制作にあたり，アンケート，インタビューという形で，次の方々にご協力をいただきました。

　　　　　海事関連企業／海事関連団体／商船系高等専門学校の卒業生

　その結果，新しい時代の海事プロフェッショナルに大切なのは次の3つがバランスよく備わっていることだということがわかってきました。

　① 海事関連の知識としての「知」
　② 英語や異文化コミュニケーション力を含む「技」
　③ 心のあり方や姿勢としての「心」

　これを図で示せば，次のようになるでしょう。

　海事業界に限らず，グローバル化が進む今日の社会では，多様な人たちと共に仕事をしていくために，英語力や異文化コミュニケーション力などの技，そして変化に富んだ状況のなかで生きていくための心のあり方や姿勢という基盤のもとに，それぞれの分野の知識が必要になってきます。

❖ 本書における英語力・コミュニケーション能力とは

　海を舞台に活躍している先輩たちの多くは，「在学中は英語が苦手だった」といいます。そんな彼らも入社して1年経つ頃には，必ずしもTOEICで高得点を獲得していなくても，立派に外国人クルーとコミュニケーションをとり大きな船を動かしています。どうやら英語力・コミュニケーション能力といっても，ひとつの物差しで測れるものではなさそうです。

　本書では，将来海を舞台に仕事をする学生の英語力・コミュニケーション能力を，海のプロフェッショナルに必要な知，技，心という視点から以下のように定義します。

海を舞台にする仕事において
海事の知識と英語の基礎力を駆使して
円滑かつ安全に仕事を遂行するために
主体的にコミュニケーションをとることができる能力

　いま，この本を手にしているみなさんは，それぞれ得意なことや強みがあると思います。英語が得意という人もいれば，「船のことならまかせてくれ！」という人もいるでしょう。また，折れない心や好奇心では誰にも負けない，という人もいると思います。海のプロフェッショナルに必要な知，技，心を統合する英語を使ったコミュニケーション能力を習得するにあたって，それぞれの強みが第一歩であり，入り口になると考えます。そこから，いまの自分に欠けているもの，これから育むべき能力を知り，伸ばしていけば，次のステップに進むことができます。

　本書では，次に示す構成を通して，海事の知識としての「知」，英語や異文化コミュニケーション能力などの習得し磨かれ続けるものとしての「技」，そして海のプロフェッショナルとしての心のあり方や姿勢としての「心」をバランスよく学んでもらえるように工夫しました。

本書の構成と利用法

　本書は，商船系高等専門学校を卒業したばかりの新米の航海士，機関士，そして無線通信士の3人を主要登場人物として，新入社員の初年度の仕事を追うストーリー仕立てとなっています。全部で20のストーリーから成り，それぞれのストーリーで海を舞台にする仕事，そしてそこで使われる英語に焦点を当てています。

✣ 構成

① **ストーリー**
　　ダイアログの背景となるストーリーを日本語で紹介する。
② **海しごと＆海事知識**
　　各ストーリーやダイアログに関係する海事用語や業務／仕事の内容を主に日本語で説明し，海事用語は日本語と英語を並記する。
③ **ダイアログ**
　　登場人物の会話。音声が用意されているので，音読練習などに使用できる。
④ **フォーカスする文法／表現**
　　ダイアログに出てくる文法や表現のなかから大切なポイントや便利な表現などを抽出。練習問題で繰り返し学ぶことができる。また，主語は網掛け，（述語）動詞はゴシック体とし，英文の基本構造を意識しやすくなるように工夫した。
⑤ **リーディング・海技士試験英語対策**
　　各ストーリーと関連のある長文の和訳問題や条約，規定事項などから抜粋。海技士試験に合格する力と知識が身につく。

✣ 音声

　ダイアログの音声は，次のURLからダウンロードできます。
　　　　http://www.kaibundo.jp/ne.htm

❖ 本書の海事英語

　本書で用いられている海事関係のフレーズの多くは，STCW条約（International Convention on Standards of Training Certification and Watchkeeping for Seafarers，船員の訓練及び資格証明並びに当直の基準に関する国際条約）で定められているSMCP（Standard Maritime Communication Phrases）に基づいています。

❖ 他の海事関連書籍と本書の位置づけ

　本書は，「海」「しごと」「英語」という3つのテーマを広くつなげることを意図したもので，何か一つに特化した専門書というよりは，総合的な「海を舞台とする仕事とそこにおけるコミュニケーション」に関する入門書，というイメージでつくられました。

　そのため，これまで出版された専門性の高い優れた海事関連／海事英語関連の教材を参照させていただきました。

　海と船を舞台とする仕事に関しては，商船高専キャリア教育研究会編『船しごと，海しごと。』を参考にしています。とくに，航海士や機関士の仕事，そして労働環境に関する記載を使用させていただきました。

　船上でのコミュニケーションに関しては，STCW条約の定めるSMCPに基づいたフレーズが多く掲載されている商船高専海事英語研究会編『はじめての船上英会話［二訂版］』を参考にさせていただきました。また，船上の作業ごとにフレーズを集めた（財）船員教育振興協会発行の「CDではじめる海の基礎英会話」も参考にさせていただきました。

　また，「リーディング・海技士試験英語対策」の題材の一部は海技士国家試験に出題された問題から選択しました。

　本書を手に取っていただいたみなさんには，各領域の知識を深め，確かなものにするために，ここで参照させていただいた教材をぜひ手にとっていただきたいと思います。

登場人物紹介

◎航(Kou)くん

航海士
ジュニアサードオフィサー
総合商社系海運会社勤務
入社1年目

◎海(Kai)くん

機関士
ジュニアサードエンジニア
総合商社系海運会社勤務
入社1年目

◎澪(Mio)さん

海岸局無線通信士(ポートラジオオペレーター)
港湾通信社勤務
入社2年目

◇KouくんとKaiくんが乗船する船

船の種類：自動車専用船
船名：The Maritime Dream
船籍：パナマ　　　　　　　　　　Flag State：Panama
総トン数：60,000 T　　　　　　　Gross Tonnage：60,000 T
載貨重量：19,000 t　　　　　　　Deadweight Tonnage：19,000 t
積載台数：乗用車6,300台（最大）　Carrying Capacity：6,300 cars
全長：200 m　　　　　　　　　　Length Overall：200 m
型幅：32 m　　　　　　　　　　　Breadth Moulded：32 m

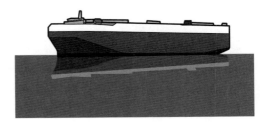

船長（Captain）：日本人
一等航海士（Chief Officer / Chief Mate）：外国人
二等航海士（Second Officer / Second Mate）：日本人
三等航海士（Third Officer / Third Mate）：外国人
次席三等航海士（Junior Third Officer / Junior Third Mate）：Kouくん

機関長（Chief Engineer）：日本人
一等機関士（First Engineer）：外国人
二等機関士（Second Engineer）：日本人
三等機関士（Third Engineer）：外国人
次席三等機関士（Junior Third Enginer）：Kaiくん

Story 1 Types of Ship We Operate

今日は新入社員研修の第1日目。航海士として仕事をしていくKouくん，機関士として仕事をしていくKaiくん共に研修に参加します。研修では先輩社員が船の種類についての概要を教えてくれるようです。

◎海しごと＆海事知識

島国である日本で生きる私たちの生活は，衣食住のための原料からエネルギーまで，海運による物資輸送に支えられています。そんな海運を担う船（vessel）にはさまざまな種類があります。

1. 石油タンカー：oil tanker
2. ケミカルタンカー：chemical tanker
3. ばら積み船：bulk carrier
4. コンテナ船：container ship
5. チップ専用船：chip carrier
6. 客船：passenger ship
7. LNG船：LNG carrier
8. 自動車専用船：
    pure car carrier (PCC)
9. 石炭専用船：bulk coal carrier

◉ダイアログ

Good morning, everyone. Today, I am going to talk about the types of ships we operate.
We operate about 4 types of ships. They are tankers, bulk carriers, pure car carriers and container ships. Tankers are designed to transport liquids, such as oil and chemicals. Bulk carriers are built to carry unpackaged materials such as coal and gravel. Pure car carriers are made to transport cars. Container ships are designed to carry goods in truck-size containers.
As you can tell, we operate various kinds of ships and they all serve important roles in our economy.

Story 1
Types of Ship We Operate

■ 語　彙

1. operate：〔動詞〕
　　運航する，運転する，操作する
2. transport：〔動詞〕輸送する，〜を運ぶ
3. liquid：〔名詞〕液体
4. chemicals：〔名詞〕薬品,薬物,化学製品
5. unpackaged：〔形容詞〕
　　梱包されていない
6. coal：〔名詞〕石炭
7. gravel：〔名詞〕砂利
8. truck-size：トラックサイズの
9. serve an important role in ～：
　　〜において重要な役割を果たす
10. economy：〔名詞〕経済

■ フォーカスする文法／表現

【英語ワザ-1】英語の文章の構成や仕組み

◆英文の構成の特徴：(1)導入，(2)具体例，(3)まとめ　という3部構成が多い。

(1)ではトピック（話題）を取り上げる。トピックが書いてある文をトピックセンテンスという。

(2)では，トピックについて具体的に詳しく説明する。

(3)では，(1)のトピックを繰り返したり，少し発展的・将来的なことに触れたりして，まとめる。

ダイアログをこの3部構成で分類すると以下のようになる。

(1) Good morning, everyone. Today, I **am going to talk** about the types of ships we **operate**.

(2) We **operate** about 4 types of ships. They **are** tankers, bulk carriers, pure car carriers and container ships. Tankers **are designed** to transport liquids, such as oil and chemicals. Bulk carriers **are built** to carry unpackaged materials such as coal and gravel. Pure car carriers **are made** to transport cars. Container ships **are designed** to carry goods in truck-size containers.

(3) As you **can tell**, we **operate** various kinds of ships and they all serve important roles in our economy.

（注）主語は網掛け，（述語）動詞は**ゴシック体**である。

練習A　**英語ワザ-1**の英文を読み，①~③の質問に答えなさい。

①トピックセンテンスを書き写しなさい。(1)のトピック(中心となる話題)を5語で抜き出しなさい。

　　🈲(トピックセンテンス)_____

　　　(トピックを5語で)_____

②トピックの具体例がいくつ書いてありますか？

　　🈲_____

③英文(3)から，(1)の内容と同じように述べている文(6語)を探し，抜き出しなさい。

　　🈲_____

練習B　以下の英文を読み，①~③の質問に答えなさい。
(題) Toolbox meeting
　A toolbox meeting is a short safety talk. It deals with a specific subject matter. It is normally delivered at the workplace, not at a training room. It should be short, within 5-10 minutes. The toolbox meeting focuses on a specific safety message.

①トピックセンテンスを書き写しなさい。また，トピックを2語で抜き出しなさい。

　　🈲(トピックセンテンス)_____

　　　(トピックを2語で)_____

②トピックに関して，「具体的」に書いてある文はいくつありますか？

答 _____

③「まとめ」の英文には，トピックセンテンスに出てくるトピックが4語で繰り返されています。それを書きなさい。

答（トピックを4語で）_____

【英語ワザ-2】後ろから名詞をさらに説明する「関係代名詞」：that, which
◆ the types of ships [(that) we operate] （名詞＋that＋主語＋動詞）

- 関係代名詞thatは，we operateと，the types of ships（先行詞）をつなぐ働きをする。また，後ろのthat以下（we operate）が，前の名詞（the types of ships）を修飾する。
- 訳すときは，先にwe operateを訳し，次に前の名詞（先行詞）（the types of ships）を訳す。

（例1）Today, I **am going to talk** about the types of ships [(that) we **operate**].

今日，私は[私たちが運航する]船の種類についてお話します。

（例2）The movie [(that) I **watched** yesterday] **was** very interesting.

[私が昨日見た]映画は面白かった。

練習C 次の下線部を訳しなさい。
① That **is** the ship [(that) we **are getting on**]!

答 _____

② The first thing [(that) I **should do** in the morning] **is** to check equipment.

＊equipment：〔名詞〕（集合的に）（船などの）装備　　＊check：〔他動詞〕検査する

答 _____

③ The data [(that) you enter] **must be** accurate.

＊accurate：〔形容詞〕正確な，的確な　⇔　inaccurate：〔形容詞〕不正確な，ずさんな

答 _____

④ The distance [that the iron core **can move**] **is adjusted** by an adjusting screw.

＊distance：〔名詞〕距離　　＊iron core：〔名詞〕鉄心
＊adjust：調整する　　＊adjusting screw：調整ネジ

答 _____

⑤ People **can pick** the cooked ingredients [that they **like**] from the pot.

＊cooked ingredient：調理ずみの材料

答 _____

⑥ There **are** a number of key documents [that you **must** always **carry** with you].

答 _____

Story 1
Types of Ship We Operate

【英語ワザ-3】受動態（受け身形）：be動詞 + Ved
◆基本形：be動詞 + Ved（動詞の過去分詞形）は，受け身（受動態）と呼ばれ，「～られる」と訳す。
助動詞（must, shall, should, mayなど）＋ be動詞 + Ved（過去分詞形）の受動態（受け身形）もよく用いられる。
例：be designed to V（動詞の原形），be made to V（動詞の原形），
　　be built to V（動詞の原形）

(例1) Tankers **are designed** (to transport liquids such as oil and chemicals).
　　　タンカーは（石油やケミカルなどの液体を輸送するために）設計されている。

(例2) Bulk carriers **are built** (to carry unpackaged materials).
　　　ばら積み船は（梱包されていない原料を運ぶために）つくられている。

(例3) Pure car carriers **are made** (to transport cars).
　　　自動車専用船は（車を輸送するために）つくられている。

練習D　日本語の意味になるように（　　　）内の語句を並べ替え，答えを書きなさい。

① Container ships **are designed** (goods, carry, in truck-size containers, to).
　コンテナ船は（トラックサイズのコンテナに入っている品物を運ぶために）設計されている。

　🖙 Container ships are designed _____.

② A valve **is designed** (to, currents of, control, gases, liquids and).
　バルブは（液体や気体の流れを制御するために）設計されている。

　🖙 A valve is designed _____.

③ Bulk carriers **are built** (to, non-packed, carry, materials).
ばら積み船は(梱包されていない原料を運ぶために)開発された。

答 Bulk carriers are built _____.

④ The Yellow Card (known, is, the Yellow Fever Card, as).
国際予防接種証明書は(黄熱病予防接種証明書として知られている)。

答 The Yellow Card _____.

⑤ Night work of seafarers under the age of 18 (prohibited, be, shall).
18歳以下の船員の夜勤は(禁止されるべきである)。

答 Night work of seafarers under the age of 18 _____.

⑥ The procedure (be, must, used, not) at the same time by multiple aircrafts or by multiple vessels.
そのやり方は多数の飛行機や船によって同時に(使用されてはいけない)。

答 The procedure _____ at the same time by multiple aircrafts or by multiple vessels.

⑦ We **must make sure** that (are, goods, in, delivered) a timely manner.
私たちは(物資が)適した方法(で配達されている)ということを確認しなければいけません。

答 We must make sure that _____ a timely manner.

⑧ Radio transmissions (made, soon, should, as, be) as possible.
無線通信はできるだけ(すみやかになされるべきである)。

答 Radio transmissions _____ as possible.

◇リーディング・海技試験英語対策

次の英文を日本文になおしなさい。

Bulk Carrier Safety

Bulk carriers were developed in the 1950s to carry large quantities of non-packed commodities such as grains, coal and iron ore. Some 5,000 bulk carriers trade around the world, providing a crucial service to world commodities' transportation.

Bulk carrier operators must be aware of the specific safety concerns related to this type of ship. Loading of cargo must be done carefully, to ensure cargo cannot shift during a voyage leading to stability problems. Large hatch covers must be watertight and secure.

(IMOホームページhttp://www.imo.org/OurWork/Safety/Regulations/Pages/BulkCarriers.aspxより)

Story 2　A List of Things to Pack

社会人としてのマナー，英語を学ぶ研修，および実務に関する研修が1〜2か月ほど本社で行われた後，いよいよ初乗船です。航海士KouくんはKosen港から社船の自動車船に乗船。機関士Kaiくんは一足遅れてシンガポールから同じ自動車船に乗ることになりました。そこで持ち物について先輩から説明を受けています。

◎海しごと＆海事知識

外航船の船員の必需品には以下のようなものがあります。

- 日本の船員手帳
  Mariner's Pocket Ledger

- パスポート
  Passport

- 海技免状
  Merchant Mariner Credential

- 黄熱病予防接種の証明書（イエローカード）
  The Yellow Card (The Yellow Fever Card)

- その他資格／証明書：危険物等取扱責任者，衛生管理者，消火作業指揮者など

◉ダイアログ

先輩 ：In about a week, you will be boarding our ship on your first job assignment.
How do you both feel?

Kou ：I feel a little nervous, but I think I am ready.

Kai ：I am excited and I can't wait!

先輩 ：Today, let's go over a list of things to pack. As a mariner, there are a number of key documents that you always need to carry with you. What do you think they are?

Kou ：A passport?

先輩 ：That's right. A passport is the most important identification document while you are overseas.

Anything else?

Kai ： A mariner's pocket ledger?

先輩： Absolutely. The ledger includes important information about you as a mariner such as insurance, paid vacation, health certificates, maritime employment history, as well as your basic information such as name, birthday, and address. This also serves as an ID.
Don't forget to keep the Yellow Card with you.

Kou ： The Yellow Card?

先輩： The Yellow Card is also known as the Yellow Fever Card. It is a document that shows that you have received vaccination against yellow fever.

語　彙

1. board：〔動詞〕乗船する
2. job assignment：任務，職務，任命
3. nervous：〔形容詞〕緊張して
4. document：〔名詞〕書類，文書
5. identification：〔名詞〕身分証明書
6. overseas：〔副詞〕海外に，外国に
7. ledger：〔名詞〕台帳，元帳
8. insurance：〔名詞〕保険
9. paid vacation：有給休暇
10. health certificates：健康診断書
11. employment history：職歴
12. yellow fever：黄熱病
13. vaccination：〔名詞〕予防接種

フォーカスする文法／表現

【英語ワザ-1】英語の文章の構成や仕組み

◆具体的に述べる：such asやfor example

- 英語の文章では，抽象的な語のあとに，such asと続けて具体的なことを述べる。抽象的な語の意味がわからなくても，such as以下が理解できれば，英文内容が推測できる。
- 英語の文章では，一般的な内容を述べたあとに，具体的な事例を述べる。そ

のときfor exampleを使う。for example以下が理解できれば，一般的・概念的・抽象的な英文の内容が，推測・予測できる。

■英語の文章では，一般的なこと，概念的なこと，抽象的なことが先に述べられる。次に，それらを具体的に詳細に説明するための英文が続くことが多い。

(例) You **will go** westbound visiting various ports (such as Singapore, Egypt, Greece, Italy and South Africa).

- various ports (さまざまな港) とは具体的にどういう港かということを (such as以下で) 地名をあげて説明している。

練習A 　　　の語句がsuch as, for example以下で，どのように具体的に述べられているか意識しなさい。辞書でわからない単語を調べ，下線の単語の意味を(　　)に書いて訳文を完成しなさい。

① Tankers **are designed** to transport liquids (such as oil and chemicals).

タンカーは(　　　　)や(　　　　)のような液体を輸送するために設計されている。

② Bulk carriers **are built** to carry unpackaged materials (such as coal and gravel).

ばら積み船は(　　　　)や(　　　　)のような梱包されていない原料を運ぶためにつくられている。

③ We **put** a lot of vegetables (such as Chinese cabbage, mushrooms, and green onions), tofu, and meat.

私たちは，(　　　　), (　　　　), (　　　　)のような多くの野菜，豆腐や肉を入れます。

④ The ledger **includes** important information about you as a mariner (such as insurance, paid vacation, health certificates and maritime employment history).

船員手帳は，（　　　　　），（　　　　　　　），（　　　　　　　），そして船員としての雇入履歴など，船員として重要な情報を（　　　　　　　）。

⑤ The ledger **includes** your basic information (such as name, birthday and address).

船員手帳は，（　　　　　），（　　　　　　　），そして（　　　　　　　）のような，あなたの基本的な情報を（　　　　　　　）。

⑥ I **relay** important information (such as weather, or traffic condition in the harbor) to the ship.

私は（　　　　　　　）あるいは港での（　　　　　　　）のような重要な情報をその船に（　　　　　　　）。

⑦ The third officers **are** in charge of emergency items (such as firefighting equipment and lifeboats).

三等航海士は（　　　　　　　）や（　　　　　　　）など非常時の備品について（　　　　　　　）。

⑧ A suitable marker, (for example, a smoke float or a radio beacon) **may be dropped** at the position.

たとえば（　　　　　　　）や（　　　　　　　）のような適切なマーカーがその場所に落とされることになっている。

⑨ Survivors in the water **should be lifted** in a horizontal position (if possible, for example, in two strops; one under the arms, the other under the knees).

海上での生存者は，可能であれば，たとえば2本の（　　　　　）を使って，片方は脇の下，もう一方は膝の下に入れて，水平の位置で（　　　　　）べきである。

【英語ワザ-2】気持ちを伝える表現
◆How do you feel? —— I am / feel / become / get ＋形容詞．
自分の気持ちは，「be動詞または動詞（feel, become, get）＋形容詞」で表現できる。
（例1）先輩：How **do** you **feel**?
　　　　　　どう感じますか？（どう思いますか？）
　　Kou：I **feel** a little nervous but I **am** ready.
　　　　　ちょっと不安ですが，覚悟はできています。

（例2）先輩：How **do** you **feel**?
　　　　　　どう感じますか？（どう思いますか？）
　　Kai：I **am** excited and I **can't wait**.
　　　　　ワクワクしていて，待ちきれません。

練習B　日本語の意味になるように（　　　）内の語句を並べ替え，答えを書きなさい。
① I **am not** good at English, so I (about, worried, keeping, am) the logbook in English.
英語が得意ではない。だから英語で航海日誌を書くのが心配です。

I am not good at English, so I ＿＿＿＿＿＿＿＿＿＿＿＿＿＿＿＿＿ the logbook in English.

② I (communicating, feel, with, nervous, about) the crew in English, but I **am** ready to go.
英語で乗組員と意思疎通を図ることについて不安に感じますが，進んで話しかけます。

I _____ the crew in English, but I am ready to go.

③ This **is** my first watchstanding, so I (excited, my, got, duty, at, first) and **woke up** early in the morning.
これが初めての当直です。だから初めての任務にワクワクして，朝早く目覚めました。

This is my first watchstanding, so I _____

_____ and woke up early in the morning.

④ I (very, to, got, hear, surprised) your voice over the VHF!
VHFを通してあなたの声を聞いてとても驚きました。

I _____ your voice over the VHF!

【英語ワザ-3】海仕事関連語句や表現
◆ serve as ～：～として役に立つ，～として職務を務める
(例) This ledger **serves** as an ID.
　　（IDとはidentificationの略で，「身分証明書」の意味）
　　この船員手帳は，IDとして役に立つ。

練習C　　下線の単語の意味を（　　）に書きなさい。辞書でわからない単語を調べなさい。

① You **serve** as the third engineer. **Check** the strainer and **clean** it if necessary.

あなたは三等機関士として（　　　　　）。ストレーナを（　　　　　）して（　　　　　）ならばそれを（　　　　　）しなさい。

② A logbook **serves** as important record (used in investigating a collision or accidents).

航海日誌は（　　　　　）や事故などを（　　　　　）するときに使われる重要な（　　　　　）として（　　　　　）。

③ I'm planning to serve yose-nabe.

私は，よせ鍋を（　　　　　）計画をしています。

＊serve：〔他動詞〕食事を提供する，客を応対する，職務を務める，（必要・目的に）かなう
⇒　service：〔名詞〕サービス業務，公益事業

◇リーディング・海技試験英語対策

次の英文を日本文になおしなさい。

International Convention on Standards of Training, Certification and Watchkeeping for Seafarers (STCW)

The 1978 STCW Convention was the first to establish basic requirements on training, certification and watchkeeping for seafarers on an international level. Previously the standards of training, certification and watchkeeping of officers and ratings were established by individual governments, usually without reference to practices in other countries. As a result standards and procedures varied widely, even though shipping is the most international of all industries. The Convention prescribes minimum standards relating to training, certification and watchkeeping for seafarers which countries are obliged to meet or exceed.

(IMOホームページより)

《語　注》
1. convention：〔名詞〕条約，協定
2. standard：〔名詞〕基準
3. certification：〔名詞〕資格，署名，認定
4. watchkeeping：〔名詞〕当直
5. seafarer：〔名詞〕船員
6. establish：〔動詞〕定める，制定する
7. requirement：〔名詞〕必要条件，要件
8. previously：〔副詞〕これまでは，以前は
9. individual government：各国政府
10. without reference to：照らし合わせることなしに，〜をかまわず
11. procedure：〔名詞〕手順
12. vary：〔動詞〕変わる，逸脱する
13. prescribe：〔動詞〕指示する，規定する
14. be obliged to：〜する義務がある

Story 3　Daily Responsibilities and Tasks

航海士KouくんはKosen港からジュニアサードオフィサーとして乗船することが決まりました。以前から航海士としての役割については聞いていましたが，思ったよりも一つの船を動かす上で大切な任務が多様なことに驚いています。

◎海しごと＆海事知識

一等航海士の仕事
- 船長を補佐する
- 船内の規律を維持する
- 航海部の長として，航海士・甲板部員を指揮・監督して甲板部の仕事を遂行する
- 荷役の責任者として，積荷・揚荷の立案，荷役の監督を行う
- 船体強度や安定性の計算を行う
- 入出航時は船首で係留作業や離岸作業の指揮をとる
- 航海中は航海当直を分担する

二等航海士の仕事
- 航海に関する仕事に携る
- 航海計器・操舵装置の保守点検・整備を行う
- 海図などの水路諸図誌の整理・保管および改補を行う
- 会社に提出する航海日誌を作成する
- 出入港の際は，船尾で係留作業や離岸作業の指揮をとる
- 航海時は航海当直を分担する
- 停泊中は荷役当直を分担する

三等航海士の仕事
- 航海日誌の記入・保管など書類仕事・記録の管理に携る
- 停泊中は毎朝夕，荷役の前後に喫水を読み，一等航海士に報告する

- 船長，上級航海士の指示による仕事に従事する
- 出入港の際は，船橋で船長の補佐にあたる
- 航海中は航海当直を分担する
- 停泊中は荷役当直を分担する
- 救命設備を整備する
- 水先人を船に迎え入れる

参照：『船しごと、海しごと。』，海の仕事.com

◉ダイアログ

先輩：You will be assigned to one of our car carrier ships as a junior third officer. You will board the ship in Kosen port, and will go westward visiting various ports such as Singapore, Egypt, Greece, Italy, and South Africa.

Kou：As a junior third officer, what exactly do I do?

先輩：That's a very good question. On this voyage, the most important thing for you to do is to learn how we operate our ships. The third officer will train you on the daily responsibilities and tasks aboard the ship.

Kou：I understand.

先輩：Learn as much as possible because on your second assignment, you will be on a car carrier as a third officer.

Kou：Could you please explain the role of third officers?

先輩：Sure. Generally, in addition to the duty as an officer to stand watch, third officers are in charge of items such as firefighting equipment, lifeboats, and various other emergency items. Third officers are also responsible for keeping a logbook. Oh, we cannot forget another important responsibility of the third officer. They must work to keep a good working relationship among the officers, engineers, and the crew.

Story 3
Daily Responsibilities and Tasks

Here, read this. This is a training record book. It talks about roles and responsibilities of deck officers on our ships.

語彙

1. be assigned to：〜に配属される，〜の任務を受ける
2. assignment：〔名詞〕任務，職務
3. board the ship：乗船する，船に乗る
4. onboard the ship：船の上で
5. go westward：西へ向かう
6. exactly：〔副詞〕正確に，具体的に
7. responsibility：〔名詞〕義務，責務，職責
8. task：〔名詞〕（課せられた）仕事，任務
9. explain：〔動詞〕説明する
10. role：〔名詞〕役割，任務，職務
11. duty：〔名詞〕義務，責務，仕事，任務
12. stand watch：航海当直に立つ
13. be in charge of：〜を担当して，〜を任されて
14. firefighting equipment：消防設備
15. lifeboat：〔名詞〕救命艇
16. emergency：〔名詞〕緊急事態，非常時
17. be responsible for：〜に対して責任がある
18. keep a logbook：航海日誌をつける

フォーカスする文法／表現

【英語ワザ-1】任務・割り当てについての表現-1
◆assignment, be assigned to 名詞
（例）He **was assigned to** a new duty.
　　彼は新しい仕事を割り当てられた。

*assign：〔動詞〕任命する，仕事を割り当てる
*be assigned to ～：～の仕事が割り当てられる，任命される

（例）He **left** Japan for his assignment in Asia.
　　　彼はアジアにおける任務におもむくために日本を出発した。
*assignment：〔名詞〕任命された任務，職，仕事の割り当て

【英語ワザ-2】責任・職責・任務についての表現-2

◆be in charge of 名詞

（例）You **are** in charge of night watch tonight.
　　　あなたは今晩，夜の当直の担当です。
*charge：〔名詞〕責任，義務
*be in charge of ～：～を担当して，～の責任（義務）がある＝be responsible for

【英語ワザ-3】責任・職責・任務についての表現-3

◆roles and responsibilities, be responsible for ＋ 名詞／動名詞

（例）Third officers **are** responsible for keeping a logbook.
　　　三等航海士は航海日誌をつける責任がある。
*responsible：〔形容詞〕～に責任のある
*responsibility：〔名詞〕～に対する責任（義務）

練習A　下線の単語の意味を（　　　）内に書きなさい。わからない単語は辞書で調べなさい。

① You **will be assigned to** one of our car carrier ships as a junior third officer.

　　あなたは（　　　　　）として我々の（　　　　　　　）のひとつに
　　（　　　　　　）でしょう。

Story 3
Daily Responsibilities and Tasks

② **Learn** as much as possible/ because on your second assignment, you **will be on** a car carrier as a third officer.

（　　　　）学びなさい。というのは2度目の（　　　　）で，あなたは三等航海士として自動車船に乗船するからです。

③ In about a week, you **will be boarding** our ship on your first job assignment.

およそ（　　　　），あなたは最初の仕事の（　　　　）で私たちの船に（　　　　）でしょう。

④ Third officers **are** in charge of items such as firefighting equipment, lifeboats, and various other emergency items.

三等航海士は（　　　　），（　　　　），そしてその他さまざまな（　　　　）の備品の（　　　　）。

⑤ You **are** responsible for making sure [that these equipments are in good conditions].

あなたはこれらの（　　　　）が（　　　　）であるということを確認する（　　　　）。

⑥ You **cannot forget** another important responsibility of third engineers.

あなたは，三等機関士のもうひとつの重要な（　　　　）を（　　　　）。

練習B 日本語を参考にして（　）に適切な単語のスペルを書き入れなさい。

① This is your first (　　　　　) on the ship.

　これがこの船での最初のあなたの任務です。

② As a third officer on this ship, I (　　　　) (　　　　　　　) for your training.

　この船の三等航海士として，私はあなたがたの訓練の責任を負っている。

③ This book talks about various roles and (　　　　　) of deck officers and engineers on our ships.

　この本は我々の船において航海士や機関士のさまざまな役割と責任について教えてくれる。

④ The third officer will (　　　　　　) you on the daily (　　　　　　) and tasks onboard the ship.

　三等航海士は，乗船中における日々の責任と仕事についてあなたを訓練するでしょう。

⑤ It is the (　　　　　) of the shipping agent to (　　　　　) a new officer or engineer to the ship.

　新人の航海士や機関士を船まで連れていくことは船舶代理店の責任です。

　＊it：ここでは仮主語なので，「それは」と訳さない。真の主語は，to 不定詞以下である。

Story 3
Daily Responsibilities and Tasks

◇リーディング・海技試験英語対策

次の英文を日本文になおしなさい。

Training

The prospective officer should be provided with a training record book to enable a comprehensive record of practical training and experience at sea to be maintained. The training record book should be laid out in such a way that it can provide detailed information about the tasks and duties which should be undertaken and the progress towards their completion. Duly completed, the record book will provide unique evidence that a structured programme of onboard training has been completed which can be taken into account in the process of evaluating competence for the issue of a certificate.

(STCW条約より)

《語 注》

1. prospective：〔形容詞〕
   将来の，見込まれる
2. be provided with ～：
   ～が与えられる，提供される
3. enable：〔動詞〕～を可能にする
4. comprehensive：〔形容詞〕
   包括的な，総合的な
5. practical training：実地研修
6. maintain：〔動詞〕
   積み重ねる，持続する，維持する
7. to lay out (be laid out)：
   明確に述べる，説明する
8. provide：〔動詞〕
   提供する，規定する
9. undertake (be undertaken)：
   請け負う，～に取り掛かる
10. completion：完成，完了
11. duly：〔副詞〕十分に，正式に
12. evidence：〔名詞〕証拠，証言
13. onboard training：船上訓練
14. take into account
    (be taken into account)：
    考慮に入れる
15. evaluate：〔動詞〕評価する
16. competence：〔名詞〕能力，適性
17. certificate：〔名詞〕
    証明書，免許状

Story 4　You Must Be ...

ジュニアサードエンジニアとして乗船が決まった機関士Kaiくんはシンガポールの港から乗船するため，人生で初めて飛行機に乗り，シンガポールでは無事に代理店の担当者のLeeさんと合流することができました。そしてコンテナターミナルまで連れて行ってもらう途中，船舶代理店での仕事について教えてもらいました。

◎海しごと＆海事知識

船舶代理店／エージェントの仕事
- 本船動静の把握
- 入港通報：検疫所，入国管理局，税関，港湾局
- 入港手続き：とん税の準備，岸壁申請，荷役手配
- 入港立ち会い：検疫，入港手続き，本船との打ち合わせ
- 停泊中における各種手配：飲料水，給油，船用金
- 寄港中の船員のケア：クルーの乗り降りの管理，寄港中の生活アドバイスなど

◉ダイアログ

Kai：Are you Mr. Lee?
Lee：Yes, you must be Kai. Welcome to Singapore.
Kai：Thank you.
Lee：How was your flight?
Kai：It was pretty good. I slept the whole way.
Lee：Excellent! You got some rest on the plane, then.
　　　Follow me. Let's head to the ship.
（In the car）
Lee：So this is your first assignment on the ship, right?
Kai：Yes.
Lee：Are you familiar with the job of an agent, like me?

Kai : I think I learned in the training, but I am not too sure.
Lee : OK, I will give you some examples of what we do.
　　　As a shipping agent, we prepare documents for the customs and harbor services.
Kai : Harbor services?
Lee : Yes. For example, we ensure a berth for the incoming ship, and arrange for a pilot and tugs if necessary.
Kai : I see.
Lee : We also communicate with both the ship owners and shippers to make sure that goods are delivered in a timely manner.
　　　And picking up a new officer like you, and taking him to the ship is an important part of our job. OK, we are here! That is the ship you are getting on!
Kai : Thank you!

語彙

1. be familiar with：
　　～に詳しい，～をよく知っている
2. agent：〔名詞〕代理店，代理業者
3. shipping agent：船舶代理店
4. prepare：〔動詞〕準備する，用意する
5. document：〔名詞〕文書，書類
6. customs：〔名詞〕税関
7. harbor service：港湾業務
8. ensure：〔動詞〕
　　確実にする，確認する
9. berth：〔名詞〕
　　バース，係留場所，錨地
10. incoming：〔形容詞〕入ってくる

フォーカスする文法／表現

【英語ワザ-1】初めてのあいさつの表現
◆Are you ～? 「あなたは～ですか？」
　You must be ～. 「あなたは～にちがいありません」

【英語ワザ-2】気持ちを伝える表現

◆How was (is) ～?

How was (is) ～?「～はどうでしたか（ですか）？」と，話題を振って会話を続けるのに便利なフレーズ。

自分の気持ちは，［主語＋be動詞（またはget，become，feelなど）＋形容詞］で伝える。

練習A　下記の会話例を参考に，名前がわからない人，名前をはっきりと覚えていない人と対面したとき，名前を確認するときの表現を見つけて，その英文を2つ下線に書きなさい。

Mr. B：Are you Mr. E?
Mr. E：Yes (I am and), you must be Mr. B. Welcome to Singapore, Mr. B.
Mr. B：Thank you, Mr. E.
Mr. E：How was your flight, Mr. B?
Mr. B：It was pretty good. I feel refreshed after taking a shower.

_____　　_____

練習B　挨拶をして，お互いの名前を確認した後，日本語では「どうぞよろしくお願いします。」といいます。その気持ちを表現している英文を見つけて，その英文を①の下線に書きなさい。次に，どのような表現で，話をつないでいきますか？ そのときの質問英文を②の下線に抜き出しなさい。

Mr. F：(Are you a) New crew?
Mr. B：Um… Yes, I am. My name is Mr. B.（名前をはっきりと覚えていない
　　　　人だったら）Are you Mr. G?
Mr. F：No, I am not. I am Mr. F. Nice to meet you, Mr. B.
Mr. B：Nice to meet you too, Mr. F. This is my first assignment on this ship,
　　　　so I am excited. By the way, how was your first assignment, Mr. F?
Mr. F：At the beginning, I was a little nervous, but now I am ready to teach
　　　　you what to do.

Story 4
You Must Be ...

答 ①「どうぞよろしくお願いします。」

_____

②話を続けていくときの質問英文

By the way, _____

練習C 解答例のように，自分の気持ちや考えを述べている英文を4つ抜き出し下線に書きなさい。

Mr. G：So, how was your first voyage, Mr. A?
Mr. A：At the beginning, I was not sure if I could survive such a long voyage. I was worried about my English.
Mr. G：Mr. A, I think you did a good job for the first time.
Mr. A：I am happy to hear that. How about you, Mr. C? How was your first voyage?
Mr. C：I was also worried about my English. But the crew members were very kind to me and tried to understand my broken English. I was often helped by the crew on the ship, and I got comfortable.

(解答例) I was not sure if I could survive such a long voyage.

答 ① _____

② _____

③ _____

④ _____

練習D 下記は，初対面の挨拶をした後のアイスブレイクの場面です。日本語を英語にして，AさんとBさんの対話文を作成しなさい。アイスブレイク(Break the ice)とは，緊張をほぐし，話のきっかけをつくるという意味です。これまでのストーリーや例文に出てきた英語表現を使って，日本語を英語にしなさい。

① Mr. A : _____ ?
　　こんにちは。Bさんですか？

② Mr. B : _____ .
　　はい，そうです。Aさんに違いありませんね。

③ Mr. A : _____ . _____ .
　　はい，私がAです。Bさん，お目にかかれてうれしいです(＝よろしくお願いします)。

④ Mr. B : _____

　　_____ ?
　　Aさん，私もお目にかかれてうれしいです(＝よろしくお願いします)。ところで，船旅はいかがでしたか？

⑤ Mr. A : It was my _____ . At the beginning,

　　I _____ . But _____ was very

　　kind to me. I _____ by them. _____ ?
　　初めての任務だったので，初めは不安でした。乗組員が親切な人たちでした。私はたびたび彼らに助けられました。あなたはどうでしたか？

⑥ Mr. B : It was the same with me. I was _____.

　　　　　But I was _____, _____ were also

　　　　　ready to talk to me. They are _____.
　　　　　私もあなたと同じでした。私は自分のへたな英語を心配していました。しかし，私は覚悟（心の準備）ができていました。乗組員も進んで私と話してくれました。彼らは私の友達です。

⑦ Mr. A : I tried _____ with them. I showed them

　　　　　my pictures and I _____.

　　　　　The pictures on the smart phone _____
　　　　　my poor English.
　　　　　私も乗組員と意思疎通を図ろうと努力しました。私は，彼らに写真を見せて，たこ焼きやすき焼きのような日本の食べ物を紹介しました。スマートフォンの写真（pictures on the smart phone）は，私の英語をしばしば助けてくれました。

【英語ワザ-3】 海仕事関連語句や表現

◆ensure [that 主語＋述語動詞] / ensure ＋名詞
　　　⇒ ～を確認／確保する，[that以下のこと]を確実にする，を確認する
◆make sure [that 主語＋述語動詞]
　　　⇒ [that以下のこと]を確かめる，確認する，きっと～する
・船舶運航にかかわる仕事では，作業手順，安全手順を確認しながら，任務遂行にあたることが多い。そのときに使われる動詞の使い方を学ぶ。
・that以下に「なにを」確認する（ensure）のか，「なにを」確かめる（make sure）のかを述べる。
・that（接続詞）以下には，必ず主語と述語動詞がある。

練習E 下線の単語の意味を（　　）に書きなさい。わからない語は辞書で意味を確認しなさい。

① We **ensure** a berth for the incoming ship, and **arrange for** a pilot and tugs if necessary.

　　私たちは（　　　　　）船の（　　　　　）を（　　　　　）する。
　　そして必要ならば、パイロットとタグボートを（　　　　　）。

② Lastly, **ensure** [that all the fire detection sensors are switched on].

　　最後に、すべての（　　　　　）にスイッチが入っていることを
　　（　　　　　）しなさい。

③ You **are** responsible for making sure [that these equipments are in good conditions].

　　あなたは、これらの（　　　　　）がよい状態にあることを（　　　　　）
　　（　　　　　）。

④ We also **communicate with** both the ship owners and shippers to make sure [that goods **are delivered** in a timely manner].

　　私たちはまた、品物が予定どおりに（　　　　　）ということを
　　（　　　　　）ために（　　　　　）と荷送人（荷主）と連絡を取る。

⑤ **Make sure** [that you understand the captain's standing order].

　　あなたが船長の（　　　　　）を理解しているということを
　　（　　　　　）。

⑥ The first thing [(that) I should do] **is** to make sure [that there are no suspicious persons or unidentified vessels in the port area].

私がするべき最初のことは，港湾区域にはあやしい人や（　　　　　）が
いないということを（　　　　　）ことです。

⑦ **Make sure** [that all crew **keep to** designated walkways within the port boundary].

すべての乗組員が確実に港湾境界内の（　　　　　）通路を守るように
（　　　　　）。

⑧ Masters **should ensure** [that a close watch **is** always **kept on** the course, speed and position of vessels].

船長は，船の（　　　　　），（　　　　　）と（　　　　　）がつねに
厳重に監視されている状態を（　　　　　）しなくてはならない。

⑨ **Ensure** [that the gland stuffing box is clean and all old packing has been removed].

パッキン押えをしたパッキン箱がきれいで，すべての古い（　　　　　）
が（　　　　　）ということを（　　　　　）。

＊gland：パッキン押え⇒回転またはピストン運動する軸とパッキン箱との間の詰め物として気体や液体漏れを防止する。

＊stuffing box：パッキン箱，スタッフィングボックス⇒回転シャフト，シリンダなどから出る所でガスや液体などが漏れないように詰め物をした筒状の箱。

◇リーディング・海技試験英語対策

次の英文を日本文になおしなさい。

Ship Agents

Ships are busiest while in port. In addition to loading and unloading cargo, there are stores to be taken, crew changes, inspections and repairs to be made. The list is long. Agents may arrange pilotage; schedule tugs and line service, provide customs brokerage and order services for the ship. In short, anything a ship needs while in port is coordinated through the agent. Ships come and go 24 hours a day, and an agent works with all stakeholders involved to ensure they remain on schedule. Agency work is exciting and dynamic, offering exposure to all parts of the marine industry.

（http://www.jobmonkey.com/maritimejobs/ship_agents.html より）

《語　注》

1. loading：荷積み，積み込み
2. unloading：荷下ろし
3. store：〔名詞〕在庫
4. inspection：〔名詞〕検査，検品
5. arrange pilotage：水先案内の手配をする
6. tug：〔名詞〕タグボート，曳船
7. line service：運航整備
8. customs brokerage：税関手続きの代行
9. order services：各種注文手配
10. stakeholder：〔名詞〕関係者
11. involved：～に関わっている
12. exposure to ～：
　　～に触れることができる，体験することができる，知ることができる

Story 5　Unmanned Engine Room

乗船当日，緊張の面持ちで船まで行くと，フィリピンクルーから笑顔で「新しいクルーか？」と英語で声をかけられ少し緊張が和らいだ新米機関士Kaiくん。オフィスにいたチーフからキャプテンを紹介された後，今後船上でのKaiくんの教育を担当しているサードエンジニアのところへと会いに行きました。そこでKaiくんは挨拶と共に早速，一日の流れについての説明を受けることになりました。

◎海しごと&海事知識

機関長（チーフエンジニア）の仕事
- 機関部の最高責任者
- 他の部署と調整を取りながら機関部全体をまとめ，安全運航・経済運航に努める
- 特別の場合以外，当直に入らない

一等機関士（ファーストエンジニア）の仕事
- 機関長不在の場合，機関長代行をする
- 航海中は主機の運転，整備を担当する
- 機関部部員への指示も行う
- 当直時間は4-8時，16-20時を担当する

二等機関士（セカンドエンジニア）の仕事
- 発電機，ボイラなどの保守・管理を行う
- 燃料油や潤滑油の管理も行う
- 当直時間は0-4時，12-16時を担当する

三等機関士（サードエンジニア）の仕事
- 電気全般，冷凍機，ポンプ類などの保守・管理を行う
- 機関日誌や消耗品の管理も行う
- 当直時間は8-12時，20-24時を担当する

（参考：『船しごと、海しごと。』）

◉ダイアログ

Crew : New crew?
Kai　 : Um... yes. I am Kai. Nice to meet you.
Crew : Come this way.
Kai　 : Thank you.

3/E 　: You must be Kai. Welcome.
Kai 　 : Thank you.
3/E 　: I am Thomas, the third engineer on this ship. I am in charge of your training. So, if you have any questions, ask me any time.
Kai 　 : Thank you.
3/E 　: OK, so before I show you around the ship, I will tell you about our daily schedule and what we do on the ship.
　　　　We usually get up at 7 am, then have breakfast. After that, we do morning exercise. From 8 am, engineers and crew get together in the engine room for about 15 minutes to go over safety precautions and what needs to be done. We call that a toolbox meeting. Soon after that, everybody gets to work. At noon, we usually take a lunch break, and after that we have a brief meeting to talk about the afternoon's work. At 3 pm we take a tea break, and from 3:30 to 5 pm, all the engineers prepare for "unmanned engine room" or "unattended engine room" at night.

語彙

1. in charge of ～：～を任されて，～を担当する，～の責任がある
2. go over ～：～を復習する（繰り返す），読み返す，チェックする
3. safety precautions：安全注意事項，安全対策

Story 5
Unmanned Engine Room

4. unmanned engine room：無人機関室 ⇒ 機関室無人化運転，Mゼロ運転
5. unattended：（機械など）操作員のいない，無人の

■ フォーカスする文法／表現

【英語ワザ-1】 前置詞（時間や時を表す）
◆時間や時を表す前置詞の意味：at, on, for, in, from ～ to ～, by

練習A　下線部の英語を日本語らしい語順で和訳しなさい。
- at「～に」：「狭い範囲の時」を表し，「ある一定の時」の1点を指す。
  ① <u>At noon</u>, we **usually take** a lunch break.

  訳 _____

  ② We **take** a tea break <u>at 3 p.m.</u>

  訳 _____

- on「～に」：「日，曜日や特定の時」を表し，何かの上にのっている状態を指す。
  ③ We **arrived** here in Spain <u>at 10:00 a.m.</u> <u>on 5th August</u>.

  訳 _____

  ④ The company **will complete** necessary repairs of our vessel <u>on the 8th of August</u>.

  訳 _____

47

- in「～に」:「月・季節・年」のように，期間(=容器)のなかに入っている状態を指す。
  ⑤ Our voyage **started** from Kosen port in 2013.

  訳 _____

  ⑥ **Does** it **snow** over there in winter?　　＊over there：そちらのほうでは

  訳 _____

- for「～の間」:期間的な範囲を表し，「出来事がどのくらい続いたか」に関心が向く。
  ⑦ We **usually have** a training on the same ship for half a year.

  訳 _____

- during「～の間」:期間のなかで起こったことを表し，「出来事がいつ起こったか」に関心が向く。
  ⑧ I **attended** an internship program during the summer for three weeks.

  訳 _____

- from ～ to ～「～から～まで」:fromは「起点から出発し離れていく」という状態。toは「ある行為がどこに到達するのか」という到達点に関心が向く。
  ⑨ From 3:30-5:00 p.m., all the engineers **prepare** for "unmanned engine room" or "unattended engine room" at night.

  訳 _____

- by「～まで」：「期限」を切ることを表し，byは「～が近い」というイメージが基本。

  byは，(期限)までにやるというデッドラインに関心が向く。

  until「～まで」は，その期限まで同じ状態でいることを表す。

  ⑩ When **I am** on a day shift, **I** usually **go** to work by 9 o'clock.

  訳 _____

  ⑪ Necessary repairs **were ordered and (were) completed** by 8th of August in Spain.

  訳 _____

  ⑫ But our vessel **had to be berthed** there until 10th of August because of unexpected troubles.

  訳 _____

【英語ワザ-2】 英語の文章の構成や仕組み

◆プロセス・順番についての表現：after that, then

順序よく説明するのによく使われる語句，表現がある。たとえば，1，2，3のような数字のほかに，first, second, third, lastlyあるいはafter that, thenなどである。安全運航のために，作業手順を明確にするのに役立つ表現。

練習B 順序よく説明するのに使われている語句に下線を引き，全文を和訳しなさい。

① We usually **get up** at 7 a.m. and then have breakfast. After that, we **do** morning exercise.

訳 _____

② First, **open** the strainer bypass valve. After that, **close** the sea water inlet and outlet valves. Then, **relieve** the pressure inside the housing and **dismantle** the cover. Lastly, **put** the cover back on and **tighten** the bolts in an alternating pattern.

　＊strainer bypass valve：ストレーナ（ろ過器），バイパス弁
　＊relieve：（タンクなどの圧力・真空状態を）一定値に調整する，和らげる，取り除く
　＊inside the housing：容器の内部にある
　＊dismantle：〔他動詞〕～をはずす
　＊put back：～を元に戻す
　＊alternate：〔形容詞〕交互の，1つおきの
　＊in an alternating pattern：対角に，交互に

訳 _____

③ 3/O：First, **look at** the vessel layout. This layout **shows** where abandon ship stations **are**. （中略）
Next, **look at** this poster. This **is called** a muster list, and it **is posted** throughout the ship.

　＊layout：〔名詞〕配置図

＊abandon ship station：総員退船部署
＊muster list：点呼名簿
＊throughout：〔前置詞〕〜のいたるところ

訳

【英語ワザ-3】語彙力アップ：接頭語
◆接頭語に注目すると，単語の意味が推測できる。
(例) 接頭語un-は「〜でない，不〜，無〜」という意味
　　　manned operation（有人の運転／稼働）⇒ unmanned operation（無人の運転）

練習C　　与えられた単語の意味を参考に，接頭語がついたときの単語の意味を空欄に書きなさい。

① able（〜する能力のある，〜できる）

　　unable（　　　　　　　　　　　　）

② acceptable（容認できる，好ましい）

　　unacceptable（　　　　　　　　　　　　　　　　　）

③ equal（平等な，均等な）

　　unequal（　　　　　　　　　　　　　　　　）

④ common（よく起こる，一般的な）

   uncommon（   ）

⑤ assigned（任務に就いて）

   unassigned（   ）

◆リーディング・海技試験英語対策

次の英文を日本語に訳しなさい。
Toolbox talk:
A toolbox talk is an informal group discussion that focuses on a particular safety issue. These tools can be used daily to promote your departments safety culture. Toolbox talks are also intended to facilitate health and safety discussions on the job site.

（http://www.ehs.harvard.edu/programs/toolbox-talksより）

《語　注》
1. informal：〔形容詞〕形式張らない，打ち解けた雰囲気の
2. particular：〔形容詞〕特定の
3. promote：〔動詞〕～の普及を促進する
4. department：〔名詞〕部，課，部門
5. safety culture：セーフティ・カルチャー（安全文化）
6. intend to ～：～することを目的としている
7. facilitate：〔動詞〕促進する，手助けする

本の紹介

『海の基礎英会話― Basic Maritime English』
練習船における海事英語訓練に係る検討会編
(財)船員教育振興協会発行

乗船中の自分を想像し,「こんな状況のときには英語では何ていったらいいのだろう？」そんな質問に応えてくれるのが『海の基礎英会話』。航海編・機関編と分かれて,さまざまな作業場面に関するフレーズを集めてあり,フレーズの検索がとても簡単にできます。また,よくある英語の間違った使い方や,船上コミュニケーションをスムーズにするために便利な表現など,作業に関する表現だけでなく,人間関係を円滑に保つためのヒントもたくさん詰まっている一冊です。

『はじめての船上英会話― English Communication for Cadet Ships』[二訂版]
商船高専海事英語研究会編
海文堂出版

船と船,船と陸間のコミュニケーションおよび船上作業の会話は,船の安全を守るために誤解を招かず正確に伝わる必要があります。そのため船と関わる仕事をする人は日常英語とは異なる特徴のある海事通信の英語を習得しなくてはなりません。『はじめての船上英会話』ではそのような海事通信用の英語を作業場面ごとに多く掲載しています。航海士または機関士を目指す学生や,乗船が決まった学生など,海事通信用の英語をきちんと勉強したい,という人に最適です。

Story 6　Establishing a Good Relationship

Kosen港から航海士Kouくんが乗船した自動車船は，シンガポールを経由し，これから約半年間エジプト，ギリシャ，イタリア，スペイン，イギリス，ドイツ，シンガポール，中国，インド，南アフリカなど世界各地を転々としながら，自動車の運搬をします。Kouくんは先輩からその間，一緒に乗船する外国人クルーに対して英語で自己紹介をするように言われました。

◎海しごと＆海事知識

多様化が進む外航船の現場。このような環境で仕事をする新人船員には，いったいどのような資質・技能が求められるのでしょうか。船員（海技者）の確保・育成に関する検討会報告では，外航海運事業者が新人船員に求める資質・技能などとして，以下を挙げています。

① 船員の資質として，船舶の機関及び操船に関する基礎的な知識・技能並びに船内業務及び船内生活への適応力・耐えうる精神力
② 海技者の資質として，海運会社の将来（経営）を自分が担う気持ち，基本的なコミュニケーション能力，基礎的な英語力，探求心，積極性，提案力，陸上勤務・外地駐在への意欲を求めている。

（船員（海技者）の確保・育成に関する検討会報告
http://www.mlit.go.jp/common/000205805.pdfより）

新しい環境や異文化への適応力ということに関して，外航船で日々仕事をする現役航海士および機関士の多くが「ユーモア」そして外国人船員とのコミュニケーションにおいて共通の話題を見つけることの大切さを挙げています。また安全航海には，船内雰囲気の融和と士気の向上が不可欠で，それを保つことが航海士の重要な役割でもあります。

Story 6
Establishing a Good Relationship

◉ダイアログ

Kou : What do you think is the key to a successful voyage?
先輩 : That's a good question. I think establishing a good relationship with the crew is the most important thing.
Kou : Establishing a good relationship with the crew....
先輩 : We cannot move this ship and do our job without them, right?
Kou : Right. I understand why it is important, but I don't know how.
先輩 : It's easy! Talk to them, and laugh with them! I will take you to their break room. Introduce yourself to them. If you can make them smile, then you are good to go!
Kou : How can I make them smile?
先輩 : Well, think of what you and they have in common.
Kou : Hi... Um my name is Kou. Please call me Kou. Nice to meet you! I am excited about my first assignment, but... I miss my girl friend in Japan!
Crew : Me too! (laughter)

語 彙

1. establish：〔動詞〕構築する
2. relationship：〔名詞〕関係，人間関係
3. have in common：共通点がある

フォーカスする文法／表現

【英語ワザ-1】 関係代名詞（what）
◆whatは，「〜もの，〜こと」（関係代名詞）として働く場合と，「何が，何を，何」（疑問詞）として働く場合がある。
（関係代名詞の例）
You **should think of** [what the crew and you **have** in common].
あなたは乗組員とあなたが共通して持っているものを考えてみるべきです。
＊whatは「〜もの」という意味で，関係代名詞として働き，what以下には主語と述語動詞がくる
（疑問詞の例）
What **is** the key to a successful voyage?
何が成功する航海のコツですか？
＊whatは「何が」という意味で，疑問詞として働く

練習A 関係代名詞what＋S（主語）＋V（述語動詞）
次の英文の和訳を書きなさい。

① I **will tell** you about our daily schedule and [what we **should do** on the ship].

訳 _____

② You **are going to do** this job as the third officer, so remember [what I **say**].

訳 _____

③ Sir, **could** you please **take** a look at [what I **wrote**]?
　＊take a look at：〜をちょっと見る

訳 _____

56

④ I **will give** you some examples of [what we **should do**].

訳 _____

練習B　疑問詞
和訳を参考に，（　）内の語を適切な語順に書き変えなさい。

① As the junior third engineer, (exactly, do, what, I, do)?
次席三等機関士として，私は正確には何をするのですか？

答 _____

② (What, think, you, do, is) the most important thing to work on a ship?
船で働くための最も重要なことは何だとあなたは思いますか？

答 _____

【英語ワザ-2】 動名詞（Ving）：名詞へトランスフォーム
◆Ving：動詞の意味を持ったまま，働き方が，名詞へとトランスフォーム
- 「動詞」を主語として使いたいとき，動詞をingの形にトランスフォーム（変身）
  ⇒動名詞
- 動名詞Ving形は，働きが名詞と同じである。主語（～することは），目的語（～することを），補語（～すること）〔be動詞のあとに付く語〕として働く。
- 前置詞（on, at, in, for, with, about, by, until, after, beforeなど）の後には名詞・代名詞が付く。だから，前置詞の後に動詞を付けたいときは，Ving（動名詞）を使う。
  （正）I am excited about <u>working</u> on a ship.
  （誤）I am excited about <u>work</u> on a ship.

練習C 下線部のみを和訳しなさい。

① Picking up a new officer and taking him to the ship **is** an important job for our company.

訳 _____

② Establishing a good relationship with the crew **is** the most important thing.

訳 _____

③ You **are** responsible for making sure that all the crew are familiar with how to use each emergency equipment.

訳 _____

④ Handing over and taking over a bridge watchkeeping **is** very important.
　＊hand over：～を(人の)管理にゆだねる，引き渡す，委譲する
　＊take over：引き継ぐ
　＊over：〔前置詞〕（越える動き，渡った位置をイメージして）～の向こう側に，～の上を越えて向こう側へ

訳 _____

⑤ **Make** correction by drawing double lines over the mistake.
　＊double lines：二重線
　＊over：〔前置詞〕（空間位置，離れて真上にいるイメージで）～を覆うように，～の上に

訳 _____

⑥ After discharging entire cargo, we **underwent** a survey by Mr. Smith, of Marine Company.

訳 _____

Story 6
Establishing a Good Relationship

【英語ワザ-3】 意見や個人的見解を聞く表現
◆What do you think ～?
- 「～はどう？」「～はどうかな？」と意見を求められることがよくある。そのとき，どう返事をするか？
- 最初は，単純な言い方（2～3語）で返答する。それができるようになったら，1～2文で返答する。

練習D 次の英文を2回，音読しなさい。終わったら，下記の終了マーク○を赤く塗りなさい。

○音読1回目終了
○音読2回目終了

① What do you think about the noise of the engine?
　エンジン音をどう思う？
　　(例1) No problem. I checked.
　　　　⇒ I checked the engine, and it doesn't have any problems.
　　(例2) Not good. I'll go and check it.
　　　　⇒ I think the noise is unusual. I'll go and check it.
② What do you think about the weather tonight?
　今晩の天気をどう思う？
　　(例1) Good. ⇒ It will be clear and fair.
　　(例2) Not so good. ⇒ We will soon have strong wind and rain.
　　　　　　　　　　　　We should prepare for the bad weather now.
③ What is the key to successful communication?
　何がうまくいくコミュニケーションのコツですか？
　　(例) Simple English. ⇒ Use simple English and use lots of gestures.
④ What do you think is the key to a successful voyage?
　何が成功する航海のコツだと思う？
　　(例) I think it's good communication.
　　　　⇒ I think (that) establishing good communication with the crew is the most important thing.

練習E　下記の質問に対する自分の考えを，最も簡単な表現と，ややハードルをあげた表現の2通りで答えなさい。

① What do you think is the best way to study English?
　最も簡単な表現（1文で）

　ややハードルをあげた表現（2文で）

② What do you think is a successful voyage?
　最も簡単な表現（1文で）

　ややハードルをあげた表現（2文で）

◇リーディング・海技試験英語対策

次の英文を日本文になおしなさい。

Crews that can talk to each other, laugh together and —importantly— joke together are likely to work safely and happily irrespective of the mix of their nationalities. The ability to communicate in a common language is the crucial factor determining the success of a multinational crew, regardless of what nationalities are on board, or how many. The more seafarers can understand each other, the more likely they are to run not just an efficient and safe ship, but a happy ship on which personal and working relationships can be built up.

(Chirea-Ungureanu & Vișan (2011). Teaching Communication Skills as a Prerequisite of the Course on "Intercultural Communication Onboard Ships". International Conference IMLA 19 Opatija 2011. http://www.pfri.uniri.hr/imla19/doc/012.pdf. より)

《語 注》

1. likely to：〜する可能性（確率）が高い
    ⇒ be likely to run：（船）を運航する可能性（確率）が高い
2. irrespective of：〜にかかわりなく，〜に関係なく
3. nationality：〔名詞〕国籍
4. ability：〔名詞〕能力
5. common language：共通語
6. crucial factor：重大な要素
7. determine：〔動詞〕決定する，確定する
8. multinational：〔形容詞〕多国籍の
9. regardless of：〜に関係なく
10. run：〔動詞〕運航する，運行する
11. efficient：〔形容詞〕効率的な

Story 7　Check the Strainer

朝のミーティング（toolbox meeting）で，サードエンジニアとKaiくんには，ストレーナーの掃除をするよう指示が出ました。機関士Kaiくんは，三等機関士から教えてもらいながら，初めての作業を進めていきます。船の上での作業ではエンジンなどの機械音で人の声が聞き取りにくいため，通常よりも大きな声でジェスチャーも織り交ぜながら会話をします。

◎海しごと&海事知識

機関士，機関長は主に機関室（エンジンルーム）および機関制御室（エンジンコントロールルーム）で作業を行います。

- 機関室
  船の大きさによって主機の出力が異なり，機関室の大きさも変わる。大型船になると3～4階建てのビルの大きさになり，エレベーターも設置されている。機関室には，主機，発電機をはじめ多くの機器があり，あたかも洋上プラント工場である。タービン船の場合は，主機を推進器（プロペラ）の代わりに発電機に接続させれば火力発電プラントと変わりない。発電機をはじめ主機以外の主要機器は2台以上設置することが義務づけられているので，トラブルが発生しても代替運転が可能で，船としての機能を維持できる。ところが主機は1台しかないので，トラブル・不具合が生じれば航行不能になる。このため主機は機関士がいちばん気を使い，いちばん大切に取り扱っている機械である。

- 機関制御室
  機関室内の温度は，赤道近くを運航するとき50℃前後にも達する。そのなかで作業する機関士にとっては，騒音を含め厳しい作業環境となる。このために機関室内空調設備を設け，主機を中心に主要機器の運転・監視が遠隔で行える機関制御室を設けている。　　　　　　　　　（『船しごと、海しごと。』より）

◉ ダイアログ

1/E : The sea water service line pressure is low. It is very likely that the strainer is clogged. Third engineer, check the strainer and clean it if necessary. Report to me when you are done.

3/E & Kai : OK, sir.

3/E : First, open the strainer bypass valve. After that, close the sea water inlet and outlet valves.

Kai : OK.

3/E : Then, relieve the pressure inside the housing and dismantle the cover.

Kai : OK, I see a lot of jellyfish and foreign material in the basket.

3/E : OK, clean that basket. When you are done, put it back and open the inlet valve slowly until the housing is full with sea water.

Kai : OK.

3/E : Lastly, put the cover back on and tighten the bolts in an alternating pattern. Don't forget to check for any leakage from the strainer cover.

(一部，㈶船員教育振興協会『CDではじめる海の基礎英会話』より)

語　彙

1. sea water service line：海水供給ライン
2. strainer：〔名詞〕ストレーナ（ろ過器）
3. bypass valve：バイパスバルブ，バイパス弁

4. inlet valve：入口弁
5. outlet valve：出口弁
6. relieve：〔動詞〕和らげる，取り除く
7. relieve the pressure：圧力を逃がす
8. housing：〔名詞〕ケーシング，容器，外皮，覆い
9. dismantle the cover：カバーを取り外す
10. jelly fish：クラゲ
11. foreign materials：異物
12. tighten：〔動詞〕締める ⇔ loosen：緩める
13. in an alternating pattern：交互に，対角に
14. leakage：〔名詞〕漏れ

フォーカスする文法／表現

【英語ワザ-1】 命令形の文

◆命令文では主語を省略する。動詞の原形で文を始める。「～しなさい。～せよ。」
- 海仕事では，伝達をすみやかに行うために，命令文で伝える。
- 命令文のまえにPleaseをつけると，依頼・お願いの気持ちを伝える。
 「～してください。」
- 命令文のまえにDon'tをつけると，禁止の命令になる。
 「～してはいけません。～するな。」
- 命令文のまえにLet'sをつけると，勧誘の気持ちを伝える。「～しましょう。」

(例1) **Check** the strainer and **clean** it [if (it is) necessary].
  ＊check：〔他動詞〕～を検査する
  ＊clean：〔他動詞〕～をきれいにする
  ＊if necessary：もし必要なら

(例2) **Report to** me [when you **are done**].
  ＊report to ～：〔自動詞〕～へ報告する
  ＊be done：済んだ，終了した

(例3) Please **call** me again after you **get** alongside.

(例4) **Don't be** nervous. You **'ll be** fine soon.

（例5）**Let's head** to the ship.

練習A　（　　　）に入る適切な動詞を下記の枠内より選び，そのスペルを記入しなさい。

First, open the strainer bypass valve. After that, (①　　　　) the sea water inlet and outlet valves. Then, (②　　　　) the pressure inside the housing and (③　　　　) the cover. Clean that basket. When you are done, put it back and open the inlet valve slowly. Lastly, (④　　　　) the cover back on and (⑤　　　　) the bolts in an alternating pattern. Don't forget to (⑥　　　　) for any leakage from the strainer cover.

＊strainer：ろ過機　　＊bypass valve：バイパス弁　　＊sea water：海水
＊inlet valve：入口弁，吸入弁　　＊outlet valve：出口弁　　＊housing：容器，外皮，覆い
＊in an alternating pattern：対角に，交互に　　＊leakage：漏れ

| relieve | put | close | check | tighten | dismantle |

練習B　次の英文は船が座礁したときの積荷の移動に向けた準備(Preparation for Cargo Transfer)に関する作業手順の一部です。わからない単語は辞書で意味を確認して，下線部を和訳しなさい。

① **Establish** contact with the lightening vessel and make a detailed plan of the proposed operation.

　貨物を受け取る船と（　　　　　　　　　　　），そして提案された作業の（　　　　　　　　　）。

② **Lay out** mooring lines, heaving lines, messengers, stoppers, fenders, etc.

　（　　　　　　　），（　　　　　　　　　），（　　　　　　　　　），
　（　　　　　　　），（　　　　　　　　）などを
　（　　　　　　　）。

③ **Have** the anchors **cleared** ready for use if they are in waters where use may be possible.

錨が使えそうな水域にあるならば，（
　　　　　　　　　　　　　　　　　　　　　　　　　　　　　　　）。

④ **Brief** the officers and crew on the operation with particular reference to the safety aspects.

安全面にとくに触れながら，作業に関して（
　　　　　　　　　　　　　　　　　　　　　　　　　　　　　　　）。

⑤ **Complete** the appropriate safety checklist if possible.

もし可能ならば，（
　　　　　　　　　　　　　　　　　　　　　　　　　　　　　　　）。

練習C　　下記の英文を1回音読してから，①，②，④，⑥を和訳しなさい。⑦は，実際に使うことを想定して自己紹介を書きなさい。⑧は，質問に対する自分の考えを書きなさい。

① **Let's go over** a list of things to pack.

訳 _____

② **Don't forget** to keep the Yellow Card with you.

訳 _____

③ Please **tell** me your ship's security level.
④ **Watch** and **take** good note of how the watch duties are handed over.

訳　_____

⑤ **Talk** to the crew and **laugh** with them.
⑥ **Ensure** [that another person **is standing** by the boiler when someone **is** inside (the boiler)].

訳　_____

⑦ **Introduce** yourself to the crew.
（英語で自己紹介）

_____

_____

⑧ **Think** of what you and the crew **have** in common.
　＊have 〜 in common：〜を共通点として持っている
（自分の考えを英語で）

We have _____ in common.

For example, _____.

【英語ワザ-2】つなぐ力がある接続詞
◆文と文をつなぐ接続詞
- 接続詞の例：and, but, so, or, when, if, because, until, as, afterなど
- 接続詞は，[主語(S)＋述語動詞(V)]と[主語(S)＋述語動詞(V)]をつなぐ働きをする。

練習D ①~⑦の英文と⑦~㋖をつないで意味の通る文になるように，（　）に⑦~㋖の記号を書きなさい。

① **Learn** as much as possible (　　　)
② **Open** the inlet valve slowly (　　　)
③ We **should conduct** emergency drills regularly (　　　)
④ Crew shore passes **must be** visible at all times (　　　)
⑤ **Sign** this letter of receipt, (　　　)
⑥ **Please feel** free to contact me (　　　)
⑦ I **will tell** you about our daily schedule, (　　　)

⑦ while the crew **are** within the port boundary.
㋑ before I **show** you around the ship.
㋒ because you **will be** on a car carrier as the third officer on your second assignment.
㋓ until emergency procedures **become** second nature.
㋔ until the housing **is** full with sea water.
㋕ if you **have** any questions or concerns regarding this matter.
㋖ if you **acknowledge** the shore access procedure.

【英語ワザ-3】語彙力アップ：接頭語や接尾語で意味を推測
◆反対語を意味する接頭語：un-, im-, in-, ab-, il-, ir-, dis-
反対語を表す接頭語を見分けられると，単語の意味を推測しやすい。
◆動詞へトランスフォームする接尾語：-ize (-ise), -en, -fy
「～にする」の意味となる。
-ize：prioritize, familiarize, publicize, urbanize, stabilize
-en：strengthen, broaden, soften, loosen, tighten
-fy：purify, humidify

練習E ①~⑩は形容詞の反対語です。左側の単語でわからないものがあったら辞書を使って調べなさい。それから右側の単語の意味を推測しなさい。各

単語の意味がわかったら，□に✓印をつけなさい．

① □ familiar　　　　　　　□ unfamiliar
② □ usual　　　　　　　　□ unusual
③ □ identified　　　　　　□ unidentified
④ □ perfect　　　　　　　□ imperfect
⑤ □ visible　　　　　　　□ invisible
⑥ □ normal　　　　　　　□ abnormal
⑦ □ legal　　　　　　　　□ illegal
⑧ □ regular　　　　　　　□ irregular
⑨ □ embark　　　　　　　□ disembark
⑩ □ mantle　　　　　　　□ dismantle

練習F　①～⑩は名詞／形容詞が動詞にトランスフォームした単語です．左側の単語でわからないものがあったら辞書を使って調べなさい．それから右側の単語の意味を推測しなさい．各単語の意味がわかったら，□に✓印をつけなさい．

① □ priority　　　　　　　□ prioritize
② □ familiar　　　　　　　□ familiarize
③ □ public　　　　　　　　□ publicize
④ □ urban　　　　　　　　□ urbanize
⑤ □ stable　　　　　　　　□ stabilize
⑥ □ strength　　　　　　　□ strengthen
⑦ □ broad　　　　　　　　□ broaden
⑧ □ soft　　　　　　　　　□ soften
⑨ □ pure　　　　　　　　□ purify
⑩ □ humid　　　　　　　□ humidify

練習G　①～⑤は動詞の反対語です．各単語や熟語の意味がわかったら，□に✓印をつけなさい．わからないものは辞書で意味を調べなさい．

① □ get on the ship　　　□ get off the ship
② □ switch on　　　　　　□ switch off

③ □ charge　　　　　　　□ discharge
④ □ understand　　　　　□ misunderstand
⑤ □ come in　　　　　　□ go out

◆リーディング・海技試験英語対策

次の英文を日本文になおしなさい。

Bolts and nuts used in a sea water piping system tend to become rusty and may be broken by twisting when removing. Especially a broken stud bolt remaining in the bed is a nuisance. Some said that they needed three days to reinstall a pump body due to a broken stud bolt which fixes the pump to the bottom plating. In a limited space, such works will be time-consuming.

（「The Best Seamanship —A Guide to English Skills」より）

《語　注》

1. bolt：〔名詞〕ボルト
2. nut：〔名詞〕留めネジ
3. tend to：〜する傾向がある
4. rusty：〔形容詞〕錆びた，錆で覆われた
5. stud bolt：植え込みボルト
6. remain：〔動詞〕残る
7. nuisance：〔名詞〕厄介者，不愉快なもの
8. reinstall：〔動詞〕再び設置する
9. pump body：ポンプの本体
10. bottom plating：底の板
11. limited space：限られた空間
12. time-consuming：時間のかかる

Story 8 Whenever We Have Newcomers

救命・消火設備の整備・管理，海賊対策用の防弾チョッキを含む防災備品の管理，避難訓練用ポスターの製作は通常，三等航海士の仕事です。乗船したばかりのKouくんはまず，この船の安全そして緊急時の対応に関してのオリエンテーションを先輩の三等航海士から受けています。

◎海しごと&海事知識

船上で想定される緊急事態の種類
- person overboard：海中転落／落水者
- unintentional flooding / rough weather：船内浸水／荒天
- abandon ship：退船
- collision：衝突
- grounding：座礁
- fire / explosion：火災／爆発
- major leakage or spillage of oil cargo：重大な原油流出
- medical emergency：急病，怪我などの医学的な非常事態
- security threat / piracy：安全に対する脅威／海賊

非常用設備 (emergency equipments)
① 救命設備 (life-saving equipment / survival equipment)：船舶が航行中に重大な火災・衝突などの事故が発生し，海上に避難せざるをえない状況になった場合，また，落水者が発生した際などに必要な設備
- EPIRB：非常用位置指示無線標識装置
- radio：無線
- flare：信号紅炎
- PFD (personal flotation device) / life jacket：救命胴衣
- life ring：救命浮環
- life raft：救命いかだ

- immersion suit：イマージョンスーツ
② 消防設備（firefighting equipment）
    - fire extinguisher：消火器
    - fire hydrant and hose：消火栓とホース
③ その他
    - bullet proof vest：防弾チョッキ
    - helmet：ヘルメット

◉ダイアログ

3/O：Whenever we have newcomers, it is the job of the safety officers on the ship to provide an orientation on emergency procedures as well as vessel safety and survival equipment. Pretty soon, you are going to do this as the third officer, so remember what I say well.

Kou：OK, sir.

3/O：First, look at the vessel layout. This shows the location of emergency equipments including EPIRBs, radios, PFDs, fire extinguishers, immersion suits, life rings, liferafts, and flares. This layout also shows where the abandon ship stations are. The safety officer on the ship is responsible for making sure that these equipments are in good condition and all the crew are familiar with how to use each piece of emergency equipment.

Kou：OK.

3/O：Next look at this poster. This is called a muster list, and it is posted throughout the ship, including the bridge, engine room, and crew accommodation areas. This list contains information on types of emergencies, instructions to follow in case of different types of emergency,

Story 8
Whenever We Have Newcomers

common muster points for all the crew, lifeboat assignments, and individual and team duties in case of emergency. It is the responsibility of the safety officer to update the list as necessary.

We conduct emergency drills regularly until emergency procedures become second nature. Next week, we will conduct an emergency drill. I will have you practice making a distress call.

語 彙

1. newcomer：〔名詞〕新人，初心者，新規参入者
2. provide：〔動詞〕提供する，準備する
3. emergency procedure：緊急時の対応
4. vessel safety：船の安全
5. survival equipment：救命装置
6. layout：〔名詞〕配置，見取り図
7. emergency equipment：非常用設備
8. abandon ship station：退船部署
9. muster list：点呼名簿
10. bridge：〔名詞〕船橋，ブリッジ
11. accommodation area：居住区
12. muster point：緊急時の点呼（集合）場所
13. emergency drill：避難訓練，防災訓練
14. become second nature：自然にできるようになる，習慣になる
15. distress call：遭難信号

フォーカスする文法／表現

【英語ワザ-1】前置詞（位置関係を表す）
◆位置関係（場所を含む）を表す前置詞の意味：at, on, in
- atは「点，〜で」⇒点を示すので，範囲は狭い，より具体的な場所を指す

(例1) I am sitting at the desk.

(例2) Measure the dimensions of the furnace at several points.

＊dimensions：寸法　＊furnace：加熱炉

- onは「何かの上に接触した状態でのっている，〜の上に」

⇒平面の上で起こっていること

(例1) I am sailing on a ship.

(例2) A quarter master was injured on the aft deck.

＊quarter master：操舵手

- inは「(広い場所で)〜の内部，容器に入っている，〜のなかに入っている」

⇒立方体のなかで起こっているイメージ

(例1) Fire in the galley.　＊galley：厨房

(例2) There was a problem in the main engine.

練習A　(　　)のなかに入る前置詞を，(at, in, on)から選び，記入しなさい。

① You will board the ship (　　) Kosen port.

② This is your first assignment (　　) the ship.

③ From 8 a.m., engineers and the crew get together (　　) the engine room for about 15 minutes.

④ I am sleeping (　　) the bed.

⑤ I am sleeping (　　) my room.

⑥ What are you going to serve (　　) the party tomorrow?

⑦ Do you see any abnormalities (　　) the main engine?

Story 8
Whenever We Have Newcomers

⑧ <u>It</u> **is** almost like working (　　　) an airport control tower.

【英語ワザ-2】 前置詞（つながり関係を示す）

◆つながり関係を示す前置詞の意味：with, without, within（≒in）

- withは「つながり」を表すので，「～を使って，～と共に，～という状態で」の意味。
- withoutは「外で，はずれて」を意味するoutが付いているので，「～というつながりなし，～なしで」の意味。
- withinは「なかで，くるまれて」を意味するinが付いているので，「時間・場所などが範囲以内に」の意味。

練習B　（　　　）のなかに入る適切な前置詞を，(with, without, within) から選び，記入しなさい。

① <u>It</u> **was** my pleasure working (　　　) you.

② <u>You</u> **tried** to communicate (　　　) the crew.

③ <u>You</u> **cannot board** a ship on a job assignment (　　　) a mariner's pocket ledger.

④ <u>We</u> **will be** in touch (　　　) you, as soon as <u>we</u> **receive** the bill of repairs.

⑤ <u>We</u> **cannot move** this ship and **do** our job (　　　) a good relationship with the crew.

⑥ <u>Crew shore passes</u> **must be** visible at all times, while <u>the crew</u> **are** (　　　) the port boundary.

⑦ **Make sure** that all the crew **keep to** designated walkways (　　　) the port boundary.

⑧ **Is** there any leakage or movement? No, everything **is** (　　　) the normal parameters, sir.

【英語ワザ-3】つなぐ力がある接続詞
◆文(S+V)と文(S+V)をつなぐ接続詞：whenever
・接続詞は，[主語(S)+述語動詞(V)]と[主語(S)+述語動詞(V)]をつなぐ働きをする。
wheneverは「〜するときはいつでも」の意味で，接続詞[when, if, because, until, as, after]の仲間。
・主語と述語動詞をつなぐ接続詞の位置には，下記の2つのパターンがある。
① [接続詞+S+V], [S+V].
② [S+V], [接続詞+S+V].

練習C　　下線部の英文を和訳しなさい。
① I **bring** pictures of my family whenever I **go** on a long voyage.

訳　_____

② **Report** to the captain whenever you **notice** any abnormalities.
＊abnormality：〔名詞〕異常

訳　_____

③ Whenever we **have** newcomers, it **is** the job of the third officer to provide an orientation on emergency procedures, as well as vessel safety and survival equipments.

* provide：〔他動詞〕~を準備する
* emergency procedure：非常時の対応
* A as well as B：BとさらにAも
* safety：〔名詞〕安全 ⇒ safe〔形容詞〕
* survival：〔名詞〕非常時用（生き残り）⇒ survive〔動詞〕
* equipment：〔名詞〕装置

訳 _____

④ Whenever a liner or piston **is replaced**, the clearances between the piston and liner or cylinder / **must be checked**.

* clearance：〔名詞〕隙間
* piston：ピストン
* liner：ライナ（摩擦や加熱を防ぐために2つの面の間に挟むもの）
* cylinder：シリンダー

訳 _____

⑤ Whenever it **is** appropriate, these tests **shall be recorded**.

訳 _____

練習D 次の意味を表す接続詞を次ページ☐欄から選び，スペルを記入しなさい。

① 「~のとき」　　　　　　　　　を表す接続詞（　　　　　　　）

② 「~する後で，~する前に」　　を表す接続詞（　　　　　　　）

③ 「~する間に，~する一方で」　を表す接続詞（　　　　　　　）

④ 条件「もし~ならば」　　　　　を表す接続詞（　　　　　　　）

⑤ 「～の場合に備えて，もし～の場合」　　を表す接続詞（　　　　　　　　）

⑥ 譲歩「～だけれども」　　　　　　　　を表す接続詞（　　　　　　　　）

⑦ 理由「～なので，なぜならば～」　　　を表す接続詞（　　　　　　　　）

⑧ 目的「～するように，～できるように」を表す接続詞（　　　　　　　　）

if, in case, because/since/as, although/though, when/as, while, after/before, so that

【英語ワザ-4】海仕事関連語句や表現

◆ in caseとin case ofの違いは？

意味は「もし～が起こった場合，もし～の場合」でほぼ同じだが，in caseは主語と動詞からなる文をその後にとり，in case ofは名詞をとる。

- in case 主語(S)＋述語動詞(V)の場合，in caseは接続詞である。
- in case of＋名詞の場合，in case ofは前置詞である。

練習E　次の下線部を和訳しなさい。

① This list **contains** information on types of emergency and instruction to follow in case of different types of emergency.

訳＿＿＿＿＿＿＿＿＿＿＿＿＿＿＿＿＿＿＿＿＿＿＿＿＿＿＿＿＿＿＿＿＿

② This list **contains** common muster point for all the crew, life boat assignment, and individual and teams duties in case of emergency.
　＊common muster point：共通集合場所

訳＿＿＿＿＿＿＿＿＿＿＿＿＿＿＿＿＿＿＿＿＿＿＿＿＿＿＿＿＿＿＿＿＿

③ Please **note** that actual situations and circumstances as mentioned above for your reference, in case (that) the shipment (should) **turn out** unsatisfactory to the consignees at the port of discharge.

＊shipment：〔名詞〕積荷
＊turn out：〔動詞〕〜だとわかる
＊satisfactory：〔形容詞〕満足な ⇔ unsatisfactory
＊consignee：〔名詞〕荷受人
＊discharge：〔名詞〕荷揚げ，荷降ろし

◇リーディング・海技試験英語対策

Muster list and emergency instructions
(1) This regulation applies to all ships.
(2) Clear instructions to be followed in the event of an emergency shall be provided for every person on board.
(3) Muster lists complying with the requirements of regulation III/53 shall be exhibited in conspicuous places throughout the ship including the navigating bridge, engine-room and crew accommodation spaces.
(4) Illustrations and instructions in appropriate languages shall be posted in passenger cabins and be conspicuously displayed at muster stations and other passenger spaces to inform passengers of:
　(a) their muster station;
　(b) the essential actions they must take in an emergency;
　(c) the method of donning lifejackets.

(SOLAS条約1992年より)

Story 9　This Is My First Watchstanding

航海士Kouくんは今夜8時から12時までの当直(パーゼロ)を，三等航海士とフィリピン人クルーの3人で担当します。はじめての当直に少し緊張しているKouくんは，早めに船橋に上がります。その前の当直を担当している一等航海士が，新米のKouくんに当直での仕事について説明をしてくれています。

◎海しごと＆海事知識

航海当直(watchkeeping)の担当者は，船長の代理としてブリッジを任され，安全な航海の責任を負っています。ここでは，その当直担当の交代に関する表現をみていきましょう。

　　当直を担当している士官：officer in charge of navigational watch /
　　　　　　　　　　　　　　officer of the watch (OOW)
　　当直を引き継ぐ士官：relieving officer
　　当直を引き渡す：hand over a watch
　　当直を引き継ぐ：take over a watch

当直担当士官(OOW)は完全に当直を引き渡すまでブリッジの責任を負っています。以下は，当直交代の際に確認されるべき項目です。

 1. the position：位置
 2. set due to current and the wind：海流と風の方向
 3. weather and visibility：天候と視程
 4. charted course：チャートコース
 5. gyro course：ジャイロコース
 6. magnetic compass course：マグネットコース
 7. speed：速度
 8. errors on the compasses：コンパスエラー
 9. status of the navigational equipment：航海計器の状態
10. the traffic in the area：航海海域の交通
11. any recognized risks：認識されているリスク／危険

Story 9
This Is My First Watchstanding

◉ダイアログ

C/O : You are early!

Kou : This is my first time standing watch, so I wanted to come early.

C/O : Good. Then, I can go over the process of taking over a navigational watch.

Kou : Thank you, sir.

C/O : Handing over and taking over a watch is very important to navigate the ship safely. Before handing over the watch, the officer on duty will brief the incoming officer on the position of the ship, current, the wind, weather, visibility, course and speed, and any information that is relevant to safe navigation. This is an important time to check these things to correct any errors. Make sure you understand the captain's standing order.

The most important job is to keep a good lookout to avoid a collision. We must detect any approaching vessels early and take appropriate actions to avoid collisions.

Kou : Yes, sir.

C/O : Here comes the third officer. Watch and take good notes of how the watch duties are handed over.

Kou : Yes, sir.

語彙

1. navigational watch：当直
2. brief：〔動詞〕手短かな指示を与える
3. captain's standing order：船長命令簿

4. keep a good lookout：よく見張る
5. avoid：〔動詞〕回避する，避ける
6. collision：〔名詞〕衝突
7. detect：〔動詞〕見つける，気づく
8. take appropriate actions：適切な行動をとる
9. take good notes：〜のノートをしっかり取る

フォーカスする文法／表現

【英語ワザ-1】 前置詞（位置関係を表す）
◆位置関係を表す前置詞overの用法：
　go over ／ take over ／ hand over a watch
位置関係を表す前置詞には，at, in, on以外にoverがある。
overは，①前置詞，②副詞，③動詞＋overと，働き方はいろいろである。
① overの意味は「覆う，円弧を描くように上方に覆いかぶさる，接触して覆う，一面に」。乗り越える「動き」も表す（③参照）。場所，年齢など「〜の上」を表す（反対語はunder）。
② overは，無線交信では「どうぞ」という意味で使われる。
③ overは，乗り越える「動き」を表す。「〜を越えて，渡って，〜を通りすぎて」というイメージから，動作を表す動詞（take, go, handなど）とともに用いられると，下記のような意味になる。
　go over：〜を点検する，〜をきれいにする，〜を渡る，〜を繰り返す
　take over：職務などを引き継ぐ
　hand over：（権限などを）明け渡る，譲渡する

◆前置詞句が修飾している単語を探せ！
前置詞句（前置詞 over, under, etc.＋名詞）は，前置詞句より前にある「名詞」あるいは「動詞」を修飾する。

Story 9
This Is My First Watchstanding

練習A　以下の下線部を訳しなさい。①〜⑤では，前置詞句が修飾する単語を□で囲んである。修飾されている単語がどれかを意識すると，どこからどう訳すのかわかりやすい。

① Please **spread** the new tablecloth over this table.

　訳 _____

② **Make** correction by drawing double lines over the mistake.

　訳 _____

③ The employment of any person under the age of 16 **shall be prohibited**.

　訳 _____

④ Night work of seafarers under the age of 18 **shall be prohibited**.

　訳 _____

⑤ Today, **let's go** over a list of things to pack.

　訳 _____

⑥ I **can go over** the process of taking over a navigational watch.

　訳 _____

⑦ Handing over and taking over a watch **is** very important to navigate the ship safely.

   訳 _____

⑧ **Watch** and **take** good note of how the watch duties **are handed over**.

   訳 _____

⑨ Engineers and crew **get together** to go over safety precautions.

   訳 _____

【英語ワザ-2】後ろから名詞をさらに説明する（関係代名詞）

◆関係代名詞（that, who）は，[that/who ＋ V]の形で，後ろから名詞（先行詞）を修飾する。

名詞 [that ／ who ＋ V]　＊that（who）は省略できない

① The officer on duty **will brief** the officer on next duty any information [that **is** relevant to safe navigation].
   関係代名詞that以下は，any information（名詞）を後ろから修飾する。
   訳す順は，that以下が先である。

② You **sounded** pretty good for somebody [who flunked every other English test in college].
   関係代名詞who以下は，somebody（名詞）を後ろから修飾する。
   訳す順は，who以下が先である。

Story 9
This Is My First Watchstanding

練習B　下線部を意識しながら以下の文を和訳しなさい。

① The officer on duty **will brief** the officer on next duty any information [that is relevant to safe navigation] .

　＊be relevant to 名詞：〔形容詞〕〜に関連性のある

訳 _____

② You **sounded** pretty good for somebody [who **flunked** every other English test in college] .

　＊flunk：〔他動詞〕（試験，単位など）を失敗する，を落とす
　＊every other：ひとつおきの
　＊for somebody：〜の人としては

訳 _____

③ The Yellow Fever card **is** a document [that **shows** your vaccination against yellow fever] .

　＊vaccination：ワクチン

訳 _____

④ I **communicate** with ships [that **come in** and **go out of** Kosen port] .

訳 _____

⑤ A legal document with any corrections **must have** a seal of the person [who **made** changes] .

訳 _____

85

【英語ワザ-3】不定詞（to＋Vの原形）：名詞・形容詞・副詞へトランスフォーム
◆to＋Vの原形（不定詞）は，動詞の意味を持ったまま，3つの働き方へトランスフォームする。
名詞へトランスフォーム：〜することは，〜することを，〜すること
形容詞へトランスフォーム：〜するための，〜するべき
副詞へトランスフォーム：〜するために

練習C　　下線部を意識しながら以下の文を和訳しなさい。不定詞が動詞の意味を持ちながら，名詞へトランスフォームしている例である。名詞は文中で「主語（〜は）」「目的語（〜を）」「補語（be動詞のあとに置かれる名詞）」の働きをする。

① **Don't forget** to keep the Yellow Card with you.

訳 _____

② **Don't forget** to check any leakage from the strainer cover.

訳 _____

③ It **is** the job of the third officer to provide an orientation on vessel safety and survival equipments.
　＊itは形式主語，訳さない

訳 _____

④ It **is** the responsibility of the safety officer to update the list as necessary.
　＊itは形式主語，訳さない

訳 _____

Story 9
This Is My First Watchstanding

⑤ The most important job **is** to keep a good lookout.

訳 _____

⑥ The first thing I should do **is** to check that there are no suspicious persons or unidentified vessels in the port.

訳 _____

練習D　下線部を意識しながら以下の文を和訳しなさい。「to + V」が，後ろから名詞をさらに説明(修飾)している。不定詞が動詞の意味を持ちながら，形容詞(名詞にかかる働きをする品詞)へトランスフォームしている例である。

① **Let's go over** a list of things to pack.

訳 _____

② This list **contains** information on types of emergency and instructions to follow.

訳 _____

③ I **have** a lot of things to catch up.

訳 _____

練習E　下線部を意識しながら以下の文を和訳しなさい。「to + V」が後ろから述語動詞にかかっている。不定詞が動詞の意味を持ち続けながら，副詞(動詞にかかる働きをする品詞)へトランスフォームしている例である。

① <u>Tankers</u> **are designed** to transport liquids such as oil and chemicals.

   訳 _____

② <u>Bulk carriers</u> **are built** to carry unpackaged materials such as coal and gravel.

   訳 _____

③ <u>Pure car carriers</u> **are made** to transport cars.

   訳 _____

④ <u>Container ships</u> **are designed** to carry goods in truck-size containers.

   訳 _____

⑤ <u>We</u> **must work** to keep a good working relationship among the officers, engineers and the crew.

   訳 _____

⑥ <u>We</u> **communicate** with both the ship owners and shippers to make sure that goods are delivered safely.

   訳 _____

⑦ A bridge watch keeping **is** very important to navigate a ship safely.

訳 _____

⑧ Handing over the watch **is** an important timing to check navigational information and to correct any errors.

訳 _____

⑨ We **must detect** any approaching vessel and take appropriate actions to avoid collision.

訳 _____

⑩ I wish I **could be** at home to celebrate my daughter's birthday.

訳 _____

◇リーディング・海技試験英語対策

次の英文を日本文になおしなさい。

Officers of the navigational watch shall be thoroughly familiar with the use of all electronic navigational aids carried, including their capabilities and limitations, and shall use each of these aids when appropriate and shall bear in mind that the echo-sounder is a valuable navigational aid.

The officer in charge of the navigational watch shall use the radar whenever restricted visibility is encountered or expected, and at all times in congested waters having due regard to its limitations.

The officer in charge of the navigational watch shall ensure that range scales employed are changed at sufficiently frequent intervals so that echoes are detected as early as possible. It shall be borne in mind that small or poor echoes may escape detection. （STCW条約より）

《語　注》

1. thoroughly：〔副詞〕完全に，十分に
2. electronic navigational aid：電子航海計器
3. capability：〔名詞〕性能，機能，能力
4. limitation：〔名詞〕限界
5. appropriate：〔形容詞〕妥当な，適切な
6. bear in mind：覚えておく
    ⇒ be borne in mind：（受動態）〜が心に留められる
7. echo-sounder：音響測深機
8. valuable：〔形容詞〕貴重な，役に立つ
9. restricted：〔形容詞〕限られた，制限された
10. encounter：〔動詞〕遭遇する，直面する
11. expect：〔動詞〕予期する
12. congested：〔形容詞〕輻輳した，密集した
13. water：〔名詞〕海域，水域
14. have due regard to 〜：〜に十分に配慮する
15. range scale：レンジスケール
16. employ：〔動詞〕用いる，使用する
17. sufficiently：〔副詞〕十分に，足りて
18. frequent intervals：頻繁な間隔（＝頻繁に）

Story 10　Charted Course Is 055 Degrees

一等航海士と話していると，先輩の三等航海士が船橋に上がってきました。実際にどうやって当直を交代するのか見ることができる機会です。このまたとない学びの機会にKouくんは必死にメモをとっています。

◎海しごと＆海事知識

船の上では，正確にコミュニケーションをとるために，さまざまな記号や決められた表現が使われています。ここでは，風と波の状況（sea state）に関する表現と天気記号についてみていきましょう。

- 風浪階級：Sea State

0	Calm (Glassy)	鏡のようになめらかである
1	Calm (Rippled)	さざ波がある
2	Smooth (Wavelets)	なめらか，小波がある
3	Slight	やや波がある
4	Moderate	かなり波がある
5	Rought	波がやや高い
6	Very Rough	波がかなり高い
7	High	相当荒れている
8	Very High	非常に荒れている
9	Phenomenal	異常な状態

- 天気記号：Meteological Symbols / Weather Symbols

b	Blue Sky	快晴
bc	Fine but Cloudy	晴れ
c	Cloudy	薄曇り
o	Overcast	曇り
z	Hazy	煙雲

r	Rainy	雨
rs	Rain and Snow	みぞれ
h	Hail	あられ, ひょう
s	Snow	雪
f	Foggy	霧
d	Drizzling	霧雨
t	Thunder	雷

(http://www.kohkun.go.jp/ship/position/enyou_mikata.html より)

◉ダイアログ

C/O：I will brief you before handing over the watch.
3/O：Yes, please.
C/O：Charted course is 055 degrees.
　　　Gyro compass course is 057 degrees.
　　　Magnetic compass course is 053 degrees.
　　　Speed through the water is 11.5 knots.
　　　Speed over the ground is 11.0 knots.
　　　There is a vessel on the same course, right astern, distance 1.5 miles.
　　　The vessel is faster than us.
　　　Weather is b.
　　　Sea state is moderate.
　　　Current is setting to NE.
　　　Wind direction is SSW.
　　　Wind force increased from 2 to 3.
　　　Atmospheric pressure is 1012.9 hPa.
　　　Dry temperature is 25.0 ℃.
　　　Wet temperature is 23.5 ℃.
　　　Seawater temperature is 24.0 ℃.

Story 10
Charted Course Is 055 Degrees

Visibility is 15 miles.

Present position is 2 cables left from the charted course line.

Distance to the next waypoint is 3 miles.

That's all.

Do you have any questions?

3/O : No. I do not have any questions.

C/O : You have the watch.

3/O : I have the watch.

（『はじめての船上英会話』［二訂版］より）

語　彙

1. charted course：チャートコース
2. speed through the water：
   対水速度
3. speed over the ground：対地速度
4. right astern：正船尾
5. sea state：風浪階級，海面状態
6. current：〔名詞〕海流，潮流
7. atmospheric pressure：気圧
8. dry temperature：乾球温度
9. wet temperature：湿球温度
10. waypoint：〔名詞〕変針点

フォーカスする文法／表現

【英語ワザ-1】海事英語での独特な数字の読み方を学ぶ

◆STCW条約にて習得が義務づけられているSMCP（Standard Maritime Communication Phrases）では，数字を読む際には基本的には，一桁ずつ数字を読む。また，メートルやマイルなど国によって異なる単位を使うものに関しては誤解を回避するために単位を読み上げる。

（例1）055 degrees：zero five five degrees

（例2）11.5 knots：one one decimal five knots ／ one one point five knots

◆例外として，操舵号令では，通常の読み方をする。

（例）15：fifteen

練習A　以下の数字および単位を上記の例のように読み方をまず書き，それから音読しなさい。

① 057 degrees　　② 11.0 knots　　③ 1012.9 hPa
④ 25.0 ℃　　　　⑤ 15 miles　　　⑥ 1400 hours

【英語ワザ-2】 位置関係を表す前置詞：from/to, on/off
◆ from ～は起点を示す：～から
◆ to ～は帰着点／目的地を示す：～へ
（例1）Wind force increased from 2 to 3.
（例2）Present position is 2 cables left from the charted course line.
（例3）Distance to the next waypoint is 3 miles.
◆ on：接している，上にあるイメージ
（例1）There is a vessel on the same course line.
　　　同航船がいます。
（例2）We are on the charted course line.
　　　コースライン上です。
◆ off：外れている，離れているイメージ
（例1）The vessel is passing two miles off Muroto Saki.
（例2）I observed Omae Saki Light House one point abaft the port beam, bearing 340 degrees, 12 miles off.

練習B　以下の英文を日本語に訳しなさい。
① Wind force increased from 2 to 3.

訳 _____

② Present position is 2 cables left from the charted course line.

訳 _____

Story 10
Charted Course Is 055 Degrees

③ Distance to the next waypoint is 3 miles.

訳 _____

④ Wind direction changed from N to NE.

訳 _____

⑤ There is a vessel on the same course, right astern, distance 1.5 miles.

訳 _____

⑥ At 1400 hours, I observed Omae Saki Light House one point abaft the port beam, bearing 340 degrees, 12 miles off.

訳 _____

練習C　日本語に合うように，空欄に適切な前置詞を入れなさい。

① 我々は東京湾に向けコースライン上を航行中だ。

　　We are (　　　) the charted course (　　　) Tokyo bay.

② 紀伊半島沖13マイルを航行中だ。

　　We are sailing 13 miles (　　　) the coast of the Kii Peninsula.

③ 次の変針点までの距離は？

　　What is the distance (　　　) the new waypoint?

④ 天気は快晴から晴れへと変わった。

　　Weather changed (　　　) b (　　　) bc.

⑤ 次の変針点まで12マイル。

　　12 miles (　　　) the next waypoint.

⑥ 左舷20度の船舶は反航船です。

　　The vessel 20 degrees on the port bow is (　　　) the opposite course.

⑦ 右舷後方の船舶は同航船です。

　　The vessel on the starboard quarter is (　　　) the same course.

⑧ 現在の位置はコースラインから2ケーブル右です。

　　Present position is 2 cables right (　　　) the charted course line.

⑨ No.2発電機を解列せよ。

　　Take No. 2 generator (　　　)-line.

⑩ No. 2発電機を同期投入せよ。

　　Put No. 2 generator (　　　)-line.

【英語ワザ-3】海事語彙：船各部の名称をまずマスターする
◆船の進行方向に向かって右はstarbord，向かって左はport
◆接頭語を知っておくと，覚えやすい。
fore-：前方の
aft-：後部の，船尾のほうにある
a-：〜の，〜へ（名詞について方向を示す）

練習D　　辞書を用いて，以下の単語を英訳しなさい。
① port / starboard
② port bow / starboard bow
④ bow / stern
⑤ astern / abaft / ahead

Story 10
Charted Course Is 055 Degrees

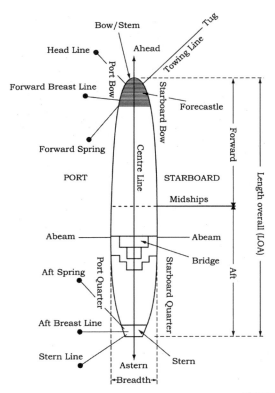

(『海事基礎英語』より)

⑥ forward / aft

⑦ midships

⑧ beam / abeam

⑨ forward breast line / aft breast line

⑩ forward spring / aft spring

⑪ forecastle

⑫ port quarter / starboard quarter

⑬ bridge

⑭ center line

⑮ breath

⑯ length overall (LOA)

◇リーディング・海技試験英語対策

次の英文を日本語になおしなさい。

As navigational and safety communications from ship to shore and vice versa, from ship to ship, and on board ship must be precise, simple and unambiguous so as to avoid confusion and error, there is a need to standardize the language used. This is of particular importance in the light of the increasing number of internationally trading vessels with crews speaking many different languages, since problems of communication may cause misunderstandings leading to dangers to the vessel, the people on board and the environment.
（中略）
Under the International Convention on Standards of Training, Certification and Watchkeeping for Seafarers, 1978, as revised 1995, the ability to use and understand the IMO SMCP is required for the certification of officers in charge of a navigational watch on ships of 500 gross tonnage or more.

（IMO STANDARD MARINE COMMUNICATION PHRASES Resolution A.918(22)より）

《語　注》

1. vice versa：逆に，逆もまた同様に
2. precise〔形容詞〕：正確な，的確な
3. unambigous〔形容詞〕：あいまいでない，はっきりした
4. avoid〔動詞〕：回避する，避ける
5. confusion〔名詞〕：混乱，取り違え
6. standardize〔動詞〕：標準化する，統一する
7. in the light of：〜を踏まえて，〜を考慮して
8. International Convention on Standards of Training, Certification and Watchkeeping：
   船員の訓練及び資格証明並びに当直の基準に関する国際条約
9. IMO (International Maritime Organization)：国際海事機関
10. SMCP (Standard Marine Communication Phrases)：標準海事通信用語集

Story 11　What Are You Going to Serve?

（乗船してから3か月が経とうとしています）
船内コミュニケーションの潤滑油となることはサードオフィサーの重要な役割の一つです。船は一人では動かせません。キャプテンから末端クルーまでが一丸となって各人の役割を果たすことが求められます。そのため，船ではクルーたちの士気を高めるため，また限られた空間に長期間滞在することの息抜きもかねて，時折，食事会やパーティーを行います。このたび航海士Kouくんは食事会の担当となりました。そこで日本の「鍋」を紹介しようと企画しています。

◎海しごと＆海事知識

船を安全に動かすには，船そして海事の専門知識や技術に代表される，業務遂行に不可欠なハードスキルが大切であることはもちろんのこと，対人関係におけるスキルや交渉力，そしてコミュニケーション力などのいわゆるソフトスキルも大切だとされています。これはヒューマンファクター（人的要因）に起因する海難事故が後を絶たないためです。2010年のマニラ改正後の「船員の訓練及び資格証明並びに当直の基準に関する国際条約（STCW）」では，「安全な航海当直の維持」の知識・理解・技能項目に「ブリッジ・リソース・マネジメント（BRM）」および「エンジンルーム・リソース・マネジメント（ERM）」が追加され，船員のチームビルディングやリーダーシップなどのソフトスキルの重要性が求められています。とくに長い航海では，クルーたちのストレスを軽減させ，またクルー内の親睦を深め，効果的なコミュニケーションや人間関係の構築に向けてさまざまな船内融和のイベントを企画するのもオフィサーの重要な役割の一つとされています。

◉ダイアログ

3/O　：What are you going to serve at the party tomorrow?
Kou　：I am planning to serve yose-nabe.

3/O : That sounds good! But most of our crew is from the Philippines. I don't think they know what a yose-nabe is. You have to explain to them what it is.

Kou : OK, I will think of ways to explain it.

（食事会開始）

Kou : Today's dinner is yose-nabe. Nabe in Japanese means pot, but it also means a soup dish cooked in nabe pot. It is often cooked at the table, and people can pick the cooked ingredients that they like from the pot.

Crew : What's in it?

Kou : We put a lot of vegetables such as Chinese cabbage, mushrooms, green onions, tofu, and meat of your choice.

Crew : How do I eat it?

Kou : You can eat it with soup or using the sauce as a dip.

Crew : This is very good. How do you say "delicious" in Japanese?

Kou : We say "Oishii."

Crew : Oishii!

Kou : Thank you, I am glad you like it!

語　彙

1. serve：〔動詞〕（飲み物，食事を）出す，提供する
2. ingredient：〔名詞〕材料

Story 11
What Are You Going to Serve?

フォーカスする文法／表現

【英語ワザ-1】 過去分詞（Ved）：形容詞へトランスフォーム
◆後ろから名詞を修飾する過去分詞（Ved）
- 名詞＋Ved（動詞の過去分詞）：動詞の意味を持ちながら，後ろから名詞を修飾（説明）する。
 （例）a soup dish cooked in nabe pot.
 名詞を修飾する形容詞へ，働き方がトランスフォーム（変身）する。
- 過去分詞1語なら，前から名詞を修飾（説明）する。
 （例）These are broken bikes.
 Ved形なので「〜られる」と受け身形のように訳す。

練習A 　下線部の「過去分詞」「過去分詞＋前置詞句」は，名詞をさらに詳しく修飾する（説明する）働きをする。下線部のVed形が修飾する名詞を（　　）に書きなさい。

① Nabe **means** a soup dish cooked in nabe pot.

　名詞（　　　　　　　　　　）

② People **can pick** the cooked ingredients [that they **like**] from the pot.

　名詞（　　　　　　　　　　）

③ What is the estimated time of arrival and departure?

　名詞（　　　　　　　　　　）

④ A logbook **serves** as important record used in investigating a collision or accidents.

　　名詞（　　　　　　　　　　　）

⑤ You **can look at** past logbook entries written by senior officers.

　　名詞（　　　　　　　　　　　）

⑥ Any corrections made on a logbook **must have** a seal of the person who made changes.

　　名詞（　　　　　　　　　　　）

⑦ All the crew **should keep to** designated walkways within the port boundary.

　　名詞（　　　　　　　　　　　）

練習B　下線部を訳しなさい。

① Nabe **means** a soup dish (cooked in nabe pot).

　　訳　_____

② People **can pick** the (cooked) ingredients [that they **like**] from the pot.

　　訳　_____

Story 11
What Are You Going to Serve?

③ What **is** the (estimated) time of arrival and departure?

訳 _____

④ A logbook **serves** as important record (used in investigating a collision or accidents).

訳 _____

⑤ You **can look at** past logbook entries (written by senior officers).

訳 _____

⑥ Any corrections (made on a logbook) **must have** a seal of the person who made changes.

訳 _____

⑦ All the crew **should keep to** (designated) walkways (within the port boundary).

訳 _____

練習C　下線部を訳しなさい。

① The crew on a ship should be familiar with the proper use of all the appliances (provided for their safety).

訳 _____

② The master should take account of any advice (given by the salvage master).
＊salvage：海難救助　＊salvage master：サルベージマスター

訳 _____

③ All the parts (used in the essential operation of the ship) should be secured against water ingress.

訳 _____

④ A cam is an eccentric projection on a revolving disk (used for the opening and closing of a valve).
＊revolving disk：回転している円盤

訳 _____

⑤ The exhaust valve housing is connected to the cylinder cover with studs and nuts (tightened by hydraulic jacks).

＊studs and nuts：植え込みボルトとナット
＊hydraulic jack：油圧ジャッキ

訳 _____

⑥ From 3:30 to 5p.m., all the engineers prepare for (unmanned) engine room or (unattended) engine room at night.

訳 _____

Story 11
What Are You Going to Serve?

【英語ワザ-2】 すでに心づもりしていることを話す表現：be going to Vの原形

◆be going to Vの原形

⇒出来上がっている計画・意図（心づもりしていること，前から〜しようと思っている）を伝える。

⇒出来事が「to Vの原形」の方向へ流れていく，向かって進んでいくこと（近い未来に起こりそうな兆候に基づいて，じきに〜しそうだ）を伝える。

練習D　　次の英文を和訳しなさい。

① What **are** you **going to serve** at the party tomorrow ?

　訳 _____

② **I am planning** to serve yose-nabe.

　訳 _____

③ **I am going to talk** about the types of ships [that we **operate**].

　訳 _____

④ Pretty soon, you **are going to do** this as the third officer, so **remember** well what **I'm saying** now.

　訳 まもなく，_____

　　 だから，_____

⑤ I think you are going to be a great engineer in a few years.

　訳 数年たつと，あなたは＿＿＿＿＿＿＿＿＿＿＿＿＿＿＿＿＿
　　　と私は思います。

【英語ワザ-3】未来を表す表現：助動詞will
◆ will＋Vの原形（未来形）
助動詞willは，単なる未来の予測「～だろう」と，その場での意思決定「～するよ」を表す。

練習E　次の下線部を和訳しなさい。
① A：The clock is broken again.
　 B：Really? I didn't notice that. OK. I'll buy a new one today.

　訳 ＿＿＿＿＿＿＿＿＿＿＿＿＿＿＿＿＿＿＿＿＿＿＿＿＿＿

② A：The clock is broken.
　 B：Yes, I know that. So I'm going to buy a new one today.

　訳 ＿＿＿＿＿＿＿＿＿＿＿＿＿＿＿＿＿＿＿＿＿＿＿＿＿＿

③ Look at the black clouds and the lightning. It is going to rain soon.

　訳 ＿＿＿＿＿＿＿＿＿＿＿＿＿＿＿＿＿＿＿＿＿＿＿＿＿＿

④ According to the weather forecast, it will rain tomorrow.

　訳 ＿＿＿＿＿＿＿＿＿＿＿＿＿＿＿＿＿＿＿＿＿＿＿＿＿＿

Story 11
What Are You Going to Serve?

⑤ You will board the ship in Kosen port as the junior third officer, and will go westbound visiting various ports.

訳 _____

⑥ We are going to board the ship to enjoy the world tour.

訳 _____

⑦ A：Can you finish the paper work by tomorrow afternoon?
　B：Yes, I will.

訳 _____

⑧ I will brief you before handing over the watch.

訳 _____

【英語ワザ-4】〇〇語では，何といいますか？　〇〇語では，△△といいます。
◆ものの名称について
　What **do** you **call** (A) in (English/Japanese/Tagalog)?
　英語／日本語／タガログ語では (A) を何と呼びますか？
　We **call** (A) [B] in (English/Japanese/Tagalog).
　英語／日本語／タガログ語では (A) を [B] といいます。
◆表現全般に使える
　How **do** you **say** (C) in English/Japanese/Tagalog?
　英語／日本語／タガログ語では (C) を何といいますか？
　We **say** (D) in English/Japanese/Tagalog.
　英語／日本語／タガログ語では (D) といいます。

◆意味について聞く

What **does** (E) **mean**?
(E)はどういう意味ですか？
(E) **means** [F] in English.
(E)は，英語では[F]という意味です。

練習F　下記の英文は，ダイアログのyose-nabeについての説明をまとめたものです。(　　　)に適切な英語を書きなさい。

① What do you call Nabe in English?

　　We call it (　　　　) in English.

② What does yose-nabe mean?

　　Yose-nabe means soup dish (　　　　) in a Nabe pot.

③ Where is yose-nabe often cooked?

　　Yose-nabe (　　　　) often (　　　　) at the table.

④ What do you put in a pot?

　　We put a lot of vegetables such as (　　　　), (　　　　), (　　　　), (　　　　) and meat of your choice.

⑤ What do you usually eat first?

　　We can pick (　　　　) (　　　　) (　　　　) that we like from the pot.

Story 11
What Are You Going to Serve?

⑥ How do you eat them?

We can eat them (　　　) soup or (　　　) it in the sauce.

（スープで食べるか，あるいはソースにちょっとつける）

練習G　以下の日本文を英文に，英文を日本文にしなさい。

① これはタガログ語では何と呼びますか？

答　_____

② Thank youはタガログ語でどういうのですか？

答　_____

③ What does "unmanned engine room" mean?

答　_____

④ How do you say "good night" in your language?

答　_____

◇リーディング・海技試験英語対策

In order to maintain a safe navigational and engineering watch, the officers of the watch must have sufficient level of knowledge of bridge / engineroom resource management principles including:

1. Allocation, assignment, and prioritization of resources

2. Effective communication
3. Assertiveness and leadership
4. Obtaining and maintaining situational awareness
5. Consideration of team experience

(一部STCW条約より)

《語　注》

1. sufficient：〔形容詞〕十分な
2. knowledge：〔名詞〕知識
3. principle：〔名詞〕原則，理念，方針
4. allocation：〔名詞〕割り当て，割り振り
5. prioritization：〔名詞〕優先順位決定
6. effective：〔形容詞〕効果的な，効率的な
7. assertiveness：〔名詞〕明確な意思表示，自分の意見を述べること
8. obtain：〔動詞〕手に入れる，得る，獲得する
9. maintain：〔動詞〕保つ，維持する
10. situational awareness：状況認識
11. consideration：〔名詞〕考慮，配慮

Story 12　I Wish I Could Be Home ...

機関士Kaiくんは一緒に働くフィリピン人クルーともっと仲良くなりたいと，時々休憩時間に彼らが集まるメスルームに顔を出します。今日はフィリピン人クルーのアンソニーがつくってくれた卵焼きを食べながら，何気ない会話を楽しんでいます。この何気ない会話こそがお互いを知る良い機会となり，日常業務も円滑に進むようになると先輩から教えてもらったKaiくん。電子辞書を駆使しながら，いろいろな話をしています。

◎海しごと＆海事知識

多くの船会社は国際競争力を保つため，船員費の高い国の船員の数を抑え，船員費の安い国の船員を雇うという仕組みで（混乗船），安定した輸送力を維持しています。現在，日本の商船に乗る外国人船員のうち約7割がフィリピン出身であるといわれています（http://www.mlit.go.jp/common/000131742.pdfによる）。外国人船員との人間関係そして仕事上での良い関係を築くには，彼らの文化に興味を持ち，理解し，そして敬意をもって接することが大切です。とくに，彼らの宗教，慣習，歴史や言語について下調べをして，相手の文化では敬遠されることや，逆に大切にされていることについて知っておくといいでしょう。そうすると，どんな配慮をもって接したらいいのかがわかるでしょう。また，家族の話やポップカルチャーなど，会話を切り出しやすいトピックをもって話しかけるのもいいでしょう。

◉ダイアログ

Anthony：Are you hungry?
Kai　　：A little...
Anthony：Here, eat this one.
Kai　　：Thank you.
Anthony：So, Kai, where in Japan are you from?

Kai	: I am from Yamagata.
Anthony	: Yamagata... Where is Yamagata?
Kai	: It is located in the northwestern part of Japan.
Anthony	: So, does it snow over there in winter?
Kai	: Yes, it snows a lot!
Anthony	: My children have never seen snow. One day, I want to take them there.
Kai	: How many children do you have?
Anthony	: I have three. Two boys and one girl. The oldest boy is 20, and the younger son is 15. My youngest daughter will turn three tomorrow.
Kai	: Tomorrow is your daughter's birthday?
Anthony	: Yes, and I wish I could be home to celebrate her birthday. This is her picture.
Kai	: She is very cute.

語彙

be located in 〜 : 〜に位置している

フォーカスする文法／表現

【英語ワザ-1】出身地を聞く表現（より具体的な場所について聞く）
◆ Where in Japan are you from?
　I'm from Toyama prefecture.

Story 12
I Wish I Could Be Home …

コミュニケーションのきっかけをつくるのに適切なトピックの一つは「ふるさと」。「出身地は，どこですか？ それはどこにありますか？」という表現を使うと，コミュニケーションのきっかけがつかめる。

練習A　次の質問に対する答えを，日本語に合うように英語で書きなさい。

① Where in Japan are you from?

　　Ans. _____.
　　私は富山出身です。

② Where in Toyama do you live?

　　Ans. _____.
　　私は射水市に住んでいます。

③ Where in Toyama is Imizu city located?

　　Ans. _____.
　　それは富山の西のほう (the western part) に位置しています。

④ What is Toyama famous for?

　　Ans. It is famous for _____.
　　おいしい水と新鮮な魚で有名です。

⑤ Where in Tokyo is your office located?

　　Ans. _____.
　　それは東京の北西部に位置しています（あります）。

練習B　ダイアログを参考に，下記の会話を英語で書きなさい。

① A : _____?
こんにちは。あなたはフィリピン (the Philippines) のどこ出身ですか？

② B : _____.
私はミンダナオ島 (Mindanao Island) 出身です。

_____.
それはマニラ (Manila) の南方に位置しています。

③ B : _____?
あなたは日本のどこ出身ですか？

④ A : _____.
私は富山出身です。

_____.
それは本州に位置していて，東京の西側にあります。

_____.
富山は，立山，水，魚などのような (such as) 自然 (nature) で有名です。

_____.
お会いできてうれしいです。

⑤ B : _____.
私もお会いできてうれしいです。

Story 12
I Wish I Could Be Home ...

_____?
そこでは，冬に雪が降りますか？

_____.
私は雪を見たことがありません。

⑥ A : _____.
はい，たくさん降ります。

_____.
この航海が終わったら，富山につれていってあげます。

⑦ B : _____
ありがとう。あなたはとてもやさしい (kind) ですね。

【英語ワザ-2】語彙力アップ（接尾語で意味や品詞を推測）

◆形容詞へトランスフォームする接尾語：
 -ern, -ive, -able, -ful, -less, -ous, -ant, -al, -y, etc.
 west〔名詞〕　　⇒ western〔形容詞〕
 expense〔名詞〕 ⇒ expensive〔形容詞〕
 danger〔名詞〕　⇒ dangerous〔形容詞〕

◆名詞へトランスフォームする接尾語：
 -tion, -ment, -ure, -ship, -th, -ance, -ity, -age, -ency, etc.
 locate〔動詞〕　　⇒ location〔名詞〕
 improve〔動詞〕　⇒ improvement〔名詞〕
 lead〔動詞〕　　　⇒ leadership〔名詞〕
 true〔形容詞〕　　⇒ truth〔名詞〕
 perform〔動詞〕　⇒ performance〔名詞〕

練習C 左側の単語でわからないものがあったら辞書を使って調べなさい。それから右側の単語の意味を推測しなさい。各単語の意味がわかったら，□に✓印をつけなさい。

① □ respect　　　　□ respectful
② □ buoy　　　　　□ buoyant
③ □ include　　　　□ inclusive
④ □ horizon　　　　□ horizontal
⑤ □ navigate　　　 □ navigable
⑥ □ storm　　　　　□ stormy
⑦ □ explode　　　　□ explosive
⑧ □ navigate　　　 □ navigational
⑨ □ pain　　　　　 □ painless
⑩ □ extinguish　　 □ extinguishable

練習D 左側の単語でわからないものがあったら辞書を使って調べなさい。それから右側の単語の意味を推測しなさい。各単語の意味がわかったら，□に✓印をつけなさい。

① □ embark　　　　□ embarkation
② □ disembark　　　□ disembarkation
③ □ maintain　　　 □ maintenance
④ □ explain　　　　□ explanation
⑤ □ visible　　　　□ visibility
⑥ □ steer　　　　　□ steerage
⑦ □ collide　　　　□ collision
⑧ □ frequent　　　 □ frequency
⑨ □ abnormal　　　 □ abnormality
⑩ □ maneuver　　　 □ maneuverability

【英語ワザ-3】 かなわない望み（仮定法）

◆仮定法：現実とかけ離れた，距離感のある状態を述べるので，時制をずらした表現を使う。

I wish I could be at home to celebrate my daughter's birthday. Look. This is her picture.

Story 12
I Wish I Could Be Home …

- wish は「かなう可能性が低い願望」や「事実と異なること」を伝える動詞である。
- 可能性のある願望には hope が使われる。

(例) **I hope** he **will come** back soon.

◆仮定法過去：現在の事実ではなく，かけ離れた状態を仮定して述べるときは「過去形」を使う。

(例1) **I wish I could be** at home.
　　「いま，家にいられたらよかったのに（実際は船上で仕事中だ）」

現実に家にいるならば I can be at home today. である。実際には家にいないので，過去形を使う。

(例2) **I wish it was** sunny now.　「いま，晴れならいいのに」

現実に晴れていれば It is sunny. 実際には晴天ではないので，過去形を使う。

◆仮定法過去完了：過去の事実ではない状態を仮定して述べるときは，さらにかけ離れた状態をイメージするので，過去完了形（had + Ved 過去分詞形）を使う。

(例) **I wish I had called** my daughter last night.
　　「昨夜，娘に電話をしておけばよかったのに」

練習E　下線部を和訳しなさい。

◆ I wish + 主語 + V過去形「いま，～ならいいなあ」

① **My computer doesn't work** swiftly. **I wish I had** a new model.

訳 _____

② A : **Can you speak** English well?
　B : Oh, no. **I wish I could be** a good speaker of English.

訳 _____

③ **We wish** there **were no earthquakes** in Japan.

訳 _____

◆主語+動詞, as if 主語+V過去形「いま，あたかも〜かのように」

④ I feel as if I had a bad dream now.

訳 _____

⑤ It seems as if she met a ghost.

訳 _____

◆If 主語+過去形, 主語+would/could/might/should+V原形
「もしいま〜ならば，〜だろう」

⑥ If I were you, I would meet the challenge.

訳 _____

⑦ If the equipment didn't work well, we should replace it into a new one.

訳 _____

⑧ If we had a north wind now, we could sail much faster.

訳 _____

◆If 主語+had+V過去分詞, 主語+would/could/might/should+have+V過去分詞「もし，あのとき〜だったならば，〜だっただろう」

⑨ If the weather had not been bad at that time, we could have avoided such a situation.

訳 _____

⑩ If **I had studied** much harder in my younger days, **I could have passed** the test then.

訳 _____

◇リーディング・海技試験英語対策

以下の英文を日本文になおしなさい。

The world merchant fleet is a global workplace and has a long tradition for sailing with crews that represent many different nationalities. Two thirds of the world merchant marine vessels have crews that are multi-national and multi-lingual (Horck, 2005). Communication and language therefore, become vital components of multinational company's ability to conduct their business adequately and play a role in global activities. The absence of communication and language skills can make the daily passing of information difficult, hereby allowing miscommunication, which can jeopardize maritime operations for all involved. (Frofoldt and Knudsen (2007). The Human Element in Maritime Accidents and Disasters — a Matter of Communication)

《語 注》

1. merchant fleet：商船隊
2. represent：〔動詞〕代表する
3. nationality：〔名詞〕国籍
4. multi-national：〔形容詞〕
   多国籍の／な
5. multi-lingual：〔形容詞〕
   多言語の／な
6. vital：〔形容詞〕極めて重要な
7. components：〔名詞〕構成要素
8. adequately：〔副詞〕
   適切に，十分に
9. absence：〔名詞〕不足，欠如
10. hereby：これによって
11. jeopardize：〔動詞〕
    危険にさらす，台無しにする

Story 13 Long Time No See!

通信社勤務のMioさんは入社から1年が経ち、ようやく仕事にも慣れてきました。入港する船とVHFで連絡をとりながら安全に船が入港できるように誘導したり、船と代理店との中継ぎをしたりと、同時にたくさんのことを行わなければならないため毎日大忙しです。そんなMioさんは、入社後はじめての長期休暇をもらい、学生時代に短期留学をしていたハワイに来ています。今日は久しぶりに再会したハワイの友達とご飯を食べにきています。

◎海しごと＆海事知識

港湾通信士としての仕事
- 業務内容
  1. 船舶の動静を把握し、その情報を必要とする関係者へ提供
  2. 港湾に関するあらゆる情報の集約
  3. 船とVHFで連絡をとり、安全な出入港ができるように航行支援
  4. 船と港湾関係者との中継ぎ
  5. データベースへの情報入力・管理　など
- いちばん忙しい時間帯は船の出入りが多くなる5:30〜9:00、そして15:00〜17:00
- 通信頻度は1日（24時間）を通して100回以上になる日がある（港によってさらに多くなる）
- 港ごとにその特徴があるため、異動するたびに港の特徴を学ぶ

◉ダイアログ

Friend : Long time no see! So good to see you!
Mio　　: Good to see you too! It has been 3 years since the last time I saw you.
Friend : Wow, that long? Well, we have a lot to catch up. So, tell me about your job. What do you do?
Mio　　: I work for a port radio company.

Friend : What do you do there? Are you a DJ or something?

Mio : No, not that kind of radio. I communicate with ships that come in and go out of Kosen port. It is almost like working at an airport control tower.

Friend : Wow. I never knew this kind of job existed. So what does your typical day look like?

Mio : Well, when I am on a day shift, I usually go to work by 9 o'clock. First thing I do is to check to make sure that there are no suspicious persons or unidentified vessels in the port area. Then, I take over the watch and start communicating with the ships.

Friend : Well, what exactly do you communicate?

Mio : For example, I ask the ship what their estimated time of arrival or departure is. I also relay important information such as weather, or traffic condition in the harbor to the ship.

Friend : Wow, you have a very important job.

語 彙

1. catch up：近況を確かめる，遅れを取り戻す，新しい情報を得る
2. port radio：ポートラジオ，港湾通信
3. control tower：管制塔
4. exist：〔動詞〕存在する
5. typical：〔形容詞〕典型的な，標準的な
6. suspicious person：不審者

7. unidentified vessel：不審船
8. estimated time：予想時刻

フォーカスする文法／表現

【英語ワザ-1】 職業について聞く表現

◆ What do you do?
直訳すると「何をなさっていますか？」というこの表現。仕事をしているということを前提にしていない問いなので，さまざまな人との会話に使える。
What do you you?
（回答例1）I am a university student. 学生をしています。
（回答例2）I work for a shipping company. 船会社に勤めています。

◆ What is your job/occupation?
これは「ご職業は何ですか？」と具体的に相手の職業を聞く質問。よって，前提として相手が仕事をしているということになる。
What is your job/occupation?
（回答例1）I am an engineer. エンジニアです。
（回答例2）I am a teacher. 教師をしています。

◆ Who do you work for?
直訳すると「誰のために働いていますか？」となる。これは，所属する会社や組織について聞く質問。
Who do you work for?
（回答例1）I work for Nippon Syosen. 日本商船に勤めています。
（回答例2）I work for my father. 父の会社／店で働いています。

◆ Where do you work?
直訳すると「どこで仕事をしていますか？」となる。つまり，仕事をしている具体的な場所＝勤務地について聞いている。

Story 13
Long Time No See!

Where do you work?
（回答例1）I work in Tokyo.　東京で仕事をしています。
（回答例2）I work in NY.　ニューヨークで働いています。

練習A　以下の日本語に合うように，(　　　　　)に適切な英語を書きなさい。

① お兄さんは何をされていますか？

　　(　　　　) (　　　　　) your bother (　　　　)?

　私の兄は日本の大きな船会社に勤めています。

　　My brother (　　　　) (　　　　　) a major shipping company in Japan.

② 妹さんは何をされていますか？

　　(　　　　) (　　　　　) your younger sister (　　　　)?

　妹は高校生です。

　　My sister (　　　　) to high school.

③ あなたのご職業は何ですか？

　　(　　　　) (　　　　　) your (　　　　)?

　機関士です。

　　I (　　　　) an (　　　　).

④ あなたはどちらの会社／組織にお勤めですか？

(　　　) do you (　　　) (　　　)?

国立の航海訓練所です。

I (　　　) for National Institute for Sea Training.

⑤ 勤務地はどちらですか？

(　　　) do you work?

横浜です。

I (　　　) (　　　) Yokohama.

【英語ワザ-2】久しぶりに会うときに使う表現
◆ Long time no see!　　久しぶり！
◆ How have you been?　（最後に会ったときからいままで）どうしてた？
◆ Good to see you.　　会えてうれしい。
久しぶりに誰かに会うときには，仕事などお互いの近況報告をすると話が弾み，コミュニケーションが深まる。そのためには「自分の仕事」について簡単に英語で説明できるとよいでしょう。

練習B　「通信士の仕事」を英語で説明
ダイアログの対話文を参考に，(　　　)に適切な英語を書きなさい。
Friend : Long time no see! How have you been? What do you do now?
Mio　 : Good to see you again.

Story 13
Long Time No See!

① I work (　　　) (　　　) (　　　) (　　　) company.

港湾無線通信を運用する会社で働いています。

② I (　　　) (　　　) (　　　) [that come in and go out of Kosen port].

Kosen港を出入港する船と連絡をとります。

③ I feel (　　　) (　　　) I (　　　) at an airport control tower. ［実際の事実と異なるので仮定法で］

あたかも空港の管制塔で働いているかのように感じます。

④ I (　　　) (　　　) (　　　) what their estimated time of (　　　) or (　　　) is.

私は予定の入港時刻と出港時刻をその船に聞きます（尋ねます）。

⑤ I (　　　) important information such as (　　　) or (　　　) (　　　) in the harbor (　　　) (　　　) (　　　).

私は，港のなかの天気や交通状況のような重要な情報を，その船へ中継します。

Friend：Wow, that sounds like an exciting job!

練習C　「三等航海士の仕事」を英語で説明

Story 3のダイアログの対話文を参考にして，(　　　)に適切な英語を書きなさい。

Friend : Good to see you. How have you been? What do you do?

Kou : Good to see you again. I've been working as a junior third officer for one year. Now I am ready to become the third officer. I will explain what my responsibilities are like.

① First, I (   ) (   ).

まず，当直の任務に立ちます。

② I am in (   ) of items such as (   ) (   ), (   ), and various other (   ) items.

私は，消火機器，救命ボート，その他のさまざまな非常時の装備に責任があります。

③ I am also (   ) for (   ) a (   ).

私はまた航海日誌を書くのにも責任があります。

④ I must work so as to (   ) a good working (   ) among (   ), (   ) and the crew.

私は，士官，機関士，部員の間で良い仕事関係を保つために，働かなければいけません。

⑤ Last but not least, I must learn (   ) (   ) (   ) our ship.

最後になりましたが（＝大事なことをひとつ言い残しましたが），私は船舶の操作方法（いかに操作するか）について学ばなければいけません。

Story 13
Long Time No See!

練習D　「機関士としての一日の流れ」を英語で説明

Story 5のダイアログの対話文を参考に，(　　)に入る適切な語を，下記より選びなさい。

Friend : Long time, no see. How have you been? What do you do now?

Kai　　: Good to see you again. I have just finished one-year training on a ship as a trainee. Now I am ready to board a ship as a junior third engineer.

Friend : That sounds great. Please tell me what your daily schedule looks like on a ship.

Kai　　: We usually (①) at 7 a.m. and then have (②). After that, we (③) morning exercise. From 8 a.m. engineers and crew (④) together in the engine room for about 15 minutes to go over (⑤) and what needs to be done. We call it a toolbox meeting. Soon after that, everybody gets to work. At noon, we usually take a lunch break, and after that, we have a brief (⑥) to talk about the afternoon's work. At 3 p.m. we take (⑦), and from 3:30 to 5:00, all the engineers (⑧) for "(⑨) (⑩) room" or unattended engine room at night.

㋐ get	㋑ do	㋒ get up	㋓ prepare
㋔ unmanned	㋕ safety precautions	㋖ tea break	
㋗ breakfast	㋘ meeting	㋙ engine	

【英語ワザ-3】現在完了形（継続）：「過去から始まり，現在まで続いていること」

◆状態の継続：現在完了形は，現在に焦点が当たっている。過去に起点を持ち，いまでも続いている出来事・状態について，現在の視点から伝えている。

◆ have/has + Ved（動詞の過去分詞形）

- 「状態の継続」をhave/has + Ved形で表現するとき，since, for, How long 〜？という語を伴うことが多い。

- 進行形にできる動詞を使って，動作が継続していることを表したいときは，

127

現在完了進行形have/has + been + Vingを使う。
（例）We have been playing soccer for two hours.

◆現在形と過去形と現在完了形（継続）の違い
- How are you? は「ごきげんよう。いまのご気分はいかが？」と聞いている。
- How were you yesterday? は過去形で聞いているので，たとえば，昨日は顔を見なかったけれど，「ご機嫌いかがでしたか？」「（あのとき）ご気分はいかがでしたか？」と，過去のことを聞いている。
- How have you been? は現在完了形で「過去を起点に現在まで続いている状態」について聞いているので，「（最後に会ったときからいままで）どうしてた？」というニュアンスとなる。

練習E　ダイアログを参考に，（　　）に適切な英語を書きなさい。

Friend：(　　　　)(　　　　)(　　　　)(　　　　).

　　　　久しぶりですね。

　　　　So good to see you!

Mio　：Good to see you, too.

　　　　It (　　　　)(　　　　) three years (　　　　) I saw you last.

　　　　最後に会って以来，3年ぶりです。

Friend：(　　　　)(　　　　)(　　　　)(　　　　)?

　　　　（3年前からいままで）（状態は）いかがですか／どうすごされてきましたか？

Story 13
Long Time No See!

Mio : Very good. Since I came back from Hawaii, I have tried various things.

I ( ) ( ) for a port radio company in Kosen port for a year.

1年間，ずっとKosen港にある港湾通信社で働いています。

How about you?

Friend : You ( ) ( ) good experiences in Japan.

あなたは日本で良い経験をずっとしてきました（経験を持った）。

I ( ) also ( ) various things.

私もさまざまなことを経験してきました。

練習F 例を参考に，継続していること（毎日でなくてもよいが，現在まで継続していること）を3つ英語で書きなさい。主語は，誰でも，何でもよい。

(例1) My family has taken our dog for a walk early in the morning for three years.

(例2) Doraemon has helped Nobita for a long time. They have been good friends.

① _____

② _____

③ _____

◇リーディング・海技試験英語対策

次の英文を日本文に訳しなさい。

The San Diego Unified Port District operates a VHF-FM radio station from Harbor Control Headquarters at Shelter Island for contacting merchant ships, port pilots, and other nearby stations. Channel 16 (156.80 MHz) is for distress, urgent, and safety messages, and for calling; channel 12 (156.60 MHz) is for port operations. The station call sign is KJC-824.

(https://www.portofsandiego.org/marintime/
check-port-and-harbor-conditons/421-bay-pilots.htmlより)

《語　注》

1. San Diego Unified Port District：サンディエゴ統一港湾地区
2. Harbor Control Headquarters：港湾管制本部
3. merchant ships：商船
4. distress：〔名詞〕救難，遭難
5. urgent：〔形容詞〕緊急の
6. station call sign：呼び出し符号／コールサイン

Story 14　A Logbook Serves as an Important Record

船の上では毎日航海日誌（ログブック）をつけます。その担当はたいていサードオフィサーであり，Kouくんもジュニアサードオフィサーとして，航海日誌を記入することになりました。過去のログブックを参考に書いてみたのですが，自信がありません。そこで，先輩サードオフィサーに目を通してもらいアドバイスをもらうことにしました。

◎海しごと＆海事知識

ほとんどの船舶では，航海日誌（Logbook）の継続的な記載が義務づけられています。ログブックは航空機でいうブラックボックスと同じく，海難事故の調査などで重要な役割を果たします。また，航海に関する情報を定期的かつ継続的に記録しておくことで，GPSなどの航海計器が万が一故障しても航海が続けられる，という大切な役割も担っています。

ログブックの記載事項
1. 出入港：entering and leaving port
2. 毎時の針路：hourly course
3. ログ（対水）とOG（対地）スピード：
    speed through the water, speed over the ground
4. 天候，海象：weather and sea state
5. 機関の使用：use of engine
6. 変針：change course
7. 船内作業：various operations on board
8. 操練：training
9. 事故や突発的な出来事：accidents or emergency
10. 時刻改正：time change
11. ヌーンレポート：noon report
12. 荷役：loading and unloading
13. 燃料・清水の補給：suppy of fuel and fresh water

14. 乗組員の交代：crew change
15. その他：other

ログブックの書き方のポイント
 1. ペンで記入する
 2. 主語を省略する
 3. 過去形で書く
 4. 時刻を必ず記入する
 5. さまざまな略語が使われる（風浪階級，天気記号など）
 6. 訂正・削除は二重線で行う

◉ダイアログ

3/O：A logbook serves as an important record used in investigating a collision or accident. It also helps the crew navigate the ship when the radio, radar or GPS fails. So, the data you enter must be true and accurate.

Kou：I am not good at English, so I am worried about keeping the logbook in English.

3/O：Don't worry. You can look at the past logbook entries written by senior officers and learn to write. How about you give it a try today, and I will give you a feedback. Practice makes perfect, right?

Kou：Yes, sir.

Story 14
A Logbook Serves as an Important Record

Day and Date: Sunday, December 7, 2014									
Hours	Miles	Sea Scale	Gyro Compass	Gyro Error	Wind Direction/ Force	Weather	Temperature Sea/ Air	Barometer	Speed through water	Speed over ground
8	3	0.57	+1°	NE/ 3	c	24/ 25	1012.9	11.5	11.0
9	3			NE/ 3	c				
10	3			NE/ 3	c				
11	3			NE/ 3	c				
12	2	0.57	+1°	NE/ 2	c	24/ 25	1011.2	11.3	11.1

Noon Report:
Latitude: Longitude:
Day's Run
True Distance Run: True Average Speed: Distance Run by Log:
Average Speed by Log: R.P.M.:
Items and Remarks
0800 Gentle br'ze & cloudy Sea slight, rounds made, all's well
0900 Practiced fire and boat station drill, tested emergency equipments and found them in good condition
1100 Put clocks ahead 1 hour for S.M.T. in Long 105-00E
1200 Sea smooth, rounds made, all's well
Signature Officer Watch: ....

Kou : Sir, could you please take a look at what I wrote?

3/O : Sure. Let's see. You have entered the course, speed and weather observation all accurately. Good. I think you spelled this word wrong. Make corrections by drawing double lines over the mistake and put your seal next to it.

Kou : My seal?

3/O : Yes, a logbook is a legal document so any corrections made on it must have a seal of the person who made changes.

Kou : I understand.

語　彙

1. investigate：〔動詞〕調査する
2. fail：〔動詞〕機能しなくなる
3. accurate：〔形容詞〕正確な，間違いのない
4. entry：〔名詞〕記入，記入事項
5. give it a try：試しにやってみる
6. practice makes perfect：習うより慣れろ（ことわざ）
7. seal：〔名詞〕印，捺印
8. legal document：法的文書

フォーカスする文法／表現

【英語ワザ-1】現在完了形：「過去の出来事が，現在はどうなった・どうなっている」

- 動作の完了：現在完了形（have/has + Ved）は，現在に焦点が当たっている。だから，「ちょうど〜したところだ」と，動作の完了を伝えることができる。「ある何か（出来事）が過去に起こった。それで，現在はどうなっている」という気持ちを伝える。

◆現在完了形（完了）と過去形の違い

- 現在完了形

  I have just finished writing the logbook.

  航海日誌を書き終えたところだ。（だから，いまは手が空いているので，手伝うよ）

- 過去形

  I finished writing the logbook.

  航海日誌を書き終えた。（過去の事実を伝えている。現在，手が空いているかどうかを問題にしていない）

Story 14
A Logbook Serves as an Important Record

練習A　下線部を意識しながら以下の文を訳しなさい。

① <u>You have already entered</u> the course, speed and weather observation all accurately.

　＊enter：〔他動詞〕記入する
　＊weather observation：〔名詞〕気象観測
　＊accurately：〔副詞〕正確に

訳 _____

② <u>The captain has just ordered</u> to send a mayday.

　＊mayday：遭難信号

訳 _____

③ <u>I have enclosed</u> our approximate damage list, sketch of damaged parts, and copies of each survey report.

　＊approximate：〔形容詞〕おおよそ
　＊damage：〔名詞〕損害
　＊damaged：〔形容詞〕損害を受けた

訳 _____

④ <u>The bill of repairs has not reached us yet</u>.

　＊bill：請求書

訳 _____

◆現在完了形の受け身（受動態）：have/has been Ved は，助動詞＋be動詞＋Ved（受け身）と同じく，海仕事に関わる作業手順を説明するのに使われる表現である。

練習B 下線部を訳しなさい。

① Ballast water **should be put** only into cargo tanks [which **have been crude oil washed**].

＊crude oil washed：原油洗浄された

訳 バラスト水は＿＿＿＿＿＿＿＿＿＿＿＿＿＿＿＿＿＿＿＿のなかにだけ入れるべきである。

② No one **should enter** confined or partially confined spaces where carbon dioxide extinguishers **have been used**.

訳 ＿＿＿＿＿＿＿＿＿＿＿＿＿＿＿＿＿＿＿＿＿＿密閉空間あるいは部分的密閉空間には誰も入ってはいけない。

③ Any compartment which **has been flooded** with carbon dioxide **must be** fully **ventilated** before entry without breathing apparatus.

訳 ＿＿＿＿＿＿＿＿＿＿＿＿＿＿＿＿＿＿＿＿どんな区画室でも，呼吸器をつけずに入室する前には，十分に換気を行わなければいけない。

④ **Ensure** (that) the gland stuffing box **is** clean and all old packing **has been removed**.

訳 パッキン箱がきれいになっていて，そして＿＿＿＿＿＿＿＿＿＿＿＿＿＿を確認しなさい。

⑤ Safety **is increased** when (it is) berthing in adverse weather conditions [provided that the required thruster capacity **has been** correctly **estimated**].

Story 14
A Logbook Serves as an Important Record

> **訳** _____
> という条件ならば，悪天候のなかで停泊していても安全性は高められる。

⑥ The gudgeon pin assembly **should be checked** for security and freedom／but (the gudgeon pin assembly) **should not be dismantled** [until about 24000 running hours **have elapsed**] [unless defects **are detected** at an earlier examination].

> **訳** ピストンピン一式は，安全と故障の有無の検査をされるべきであるが，初期検査で欠陥が検出されないならば，_____，ピストンピン一式は分解されるべきではない。

【英語ワザ-2】提案の表現：How about ～？／What about ～？
◆ How (What) about ＋ 名詞（Ving動名詞）？
- 提案・勧誘：「～してはどうですか？」という気持ちを表す。この表現で，相手の行動を誘う。
- 意見を尋ねる：「～についてどう思うか」「～はどうしますか？」と，相手の意見・説明を求める。

練習C　下線部を訳しなさい。

① A：I am worried about keeping the logbook in English, because I am not good at English.
　G：Don't worry. You can look at past logbook entries written by senior officers. <u>How about giving it a try today?</u>
　A：Yes, sir.

> **訳** _____

② F : Let's check operating oil tank levels.
　　B : They are full, sir.
　　F : Good. How about cooling water tanks?
　　B : Full, sir.

訳 _____

練習D　Story 6のダイアログ(p.55)を参考にして，(　　)内を適切な語順にして，解答欄に書きなさい。
Kou　：I don't know how to establish a good relationship with the crew.
先輩：It's easy. Talk to them, and laugh with them.
Kou　：① How can I (them, smile, make, laugh, or)?
　　　　どうやって彼らを笑わせたり，にっこりさせたりできるのだろうか？
先輩：② (about, pictures, showing, them, how) of your family or your friends?
　　　　彼らにあなたの家族や友達の写真を見せてはどうですか？

解 ① How can I _____?

　　② _____ of your family or your friends?

練習E　(　　)内を適切な語順にして，解答欄に書きなさい。
A：① (are, in Japan, from, where, you)?
　　あなたは日本のどこ出身ですか？
E：I'm from Toyama. Toyama is located in the other side of Tokyo.
A：② I (Toyama, never, been, have, to).
　　私は富山に行ったことがありません。
E：③ (there, about, how, going) together when we get off the ship?
　　そこへ一緒に行くのはどうですか？

解 ① _____?

Story 14
A Logbook Serves as an Important Record

② I _____.

③ _____ together when we get off the ship?

【英語ワザ-3】お願いの表現：Could you ～ ?
◆丁寧な依頼：Could you ＋動詞?
「Please ＋命令文」より，丁寧にお願いする気持ちが伝わる。

練習F （　）に入る適切な語を下記より選びなさい。

B：Could you please (①) the roles of the third engineer?
F：No problem. Take notes.

A：Could you please show me what to do on the ship?
D：No problem. You must (②) attention to me carefully.

D：I will have you practice (③) a distress call. Remember what I am (④) to say.
A：Could you please (⑤) that phrase?
D：(⑥) repeating after me? Say it clearly and loudly.

| how about, pay, repeat, explain, making, going |

◇リーディング・海技試験英語対策

以下の英文を日本語になおしなさい。

Regulation V/28 of the 1974 SOLAS Convention, as amended, requires all ships engaged on international voyages to keep on board a record of navigational activities and incidents which are of importance to safety of navigation and which must contain sufficient detail to restore a complete record of the voyage, taking into account the recommendations adopted by the Organization. This resolution aims at providing guidance for the recording of such events:

1) Recording of information related to navigation

In addition to national requirements, it is recommended that the following events and items, as appropriate, be among those recorded:

1.1) before commencing the voyage

Details of all data relating to the general condition of the ship should be acknowledged and recorded, such as manning and provisioning, cargo aboard, draught, result of stability/stress checks when conducted, inspections of controls, the steering gear and navigational and radiocommunication equipment.

1.2) during the voyage

Details related to the voyage should be recorded, such as courses steered and distances sailed, position fixings, weather and sea conditions, changes to the voyage plan, details of pilots' embarkation/disembarkation, entry into areas covered by, and compliance with, routeing schemes or reporting systems.

1.3) on special events

Details on special events should be recorded, such as death and injuries among crew and passengers, malfunctions of shipboard equipment and aids to navigation, potentially hazardous situations, emergencies and distress messages received.

1.4) when the ship is at anchor or in a port

Details on operational or administrative matters and details related to the safety and security of the ship should be recorded.

(http://www.imo.org/blast/blastDataHelper.asp?data_id=24564&filename=A916(22).pdfより))

《語　注》

1. as amended：〜で改正／修正されたように
2. incident：〔名詞〕インシデント，ヒヤリハット，出来事
3. contain：〔動詞〕含む
4. sufficient detail：十分な詳細
5. restore：〔動詞〕再構築する，復元する
6. adopt：〔動詞〕採択する
7. the Organization：組織（この場合はIMOを指している）
8. resolution：決議
9. acknowledge：〔動詞〕承認する
10. manning：〔名詞〕船員配乗
11. provisioning：（食料などの）積み込み
12. aboard：〔動詞〕船に乗って
13. draught：〔名詞〕= draft，喫水
14. control(s)：〔名詞〕制御機器
15. steering gear：〔名詞〕操舵装置
16. embarkation：〔名詞〕乗船
17. disembarkation：〔名詞〕下船
18. compliance with：〜にのっとり，〜を順守して
19. routing scheme：航路指定方式
20. reporting system：報告システム
21. injury：〔名詞〕怪我
22. malfunction：〔名詞〕故障，異常，誤動作
23. shipboard：〔形容詞〕船上の
24. potentially：〔副詞〕潜在的に，〜の可能性がある
25. hazardous：〔形容詞〕危険な，有害な
26. operational：〔形容詞〕運航上の，操作上の
27. administrative：〔形容詞〕管理上の

Story 15　Sending a Mayday Signal

今日は，定期的に行われる緊急時を想定した訓練の日です。先週先輩から言われたとおり，Kouくんは遭難信号（distress call）の練習をします。船上で火災が発生し，消火活動が難航しているという想定です。さて，Kouくんは無事に英語で，遭難信号を送ることができるでしょうか。

◎海しごと＆海事知識

遭難信号（distress call）は，人間，航空機，船舶が重大で差し迫った危機にあり（imminent danger），即時の救助（immediate assistance）を必要としているときに発信されます。そのような状況の例としては，火災（fire），沈没（sinking），爆発（explosion）などが挙げられます。3回繰り返される「メーデー，メーデー，メーデー」は遭難信号を発信するときに国際的に使われる緊急用符号語。このメーデー呼び出しを受信したら，救助の支援となる通信以外は一切の通信が許されません。救助を求める船舶などは，メーデーの呼び出しの後に，以下の情報を発信します。

1. Who you are：船舶名，コールサイン
2. Where you are：位置（緯度，経度，方位，距離など）
3. What is happening：何が起こっているのか
4. What kind of assistance：どのような救援を必要としているのか
5. How many people：何人乗船していて，何人負傷しているのか
6. Seaworthiness：耐航性，堪航性
7. Description of your vessel：
　　船の説明（長さ，タイプ，幅，構造，船体の色など）
8. Radio Frequency：通信に使用する無線の周波数
9. Survival equipment available：利用可能な救命装備

　　　　　　　（参考 http://www.uscg.mil/d1/prevention/NavInfo/navinfo/documents/
　　　　　　　　　　　　　　　　　　　　A-Emergency_Procedures.PDF）

嘘のメーデー呼び出しは多くの国で犯罪とされています。「即刻の救助を必

要とするような重大で差し迫った脅威」(grave and imminent danger and requests immediate assistance)ではないが，機械の故障や病人の発生などの緊急(urgent)事態を知らせる場合は，「パン，パン，パン」という緊急信号を使用します。

◉ダイアログ

3/O：The captain has just ordered to send a mayday.

Kou：Roger sir. Sending a mayday signal. Switching to VHF channel 16.
This is a drill. MAYDAY, MAYDAY, MAYDAY. This is the Maritime Dream, Maritime Dream, Maritime Dream. This is a drill, over.

(Release microphone and listen for response)

Kou：MAYDAY, This is the Maritime Dream.
My position is 21° 15′S, 71° 10′E, drifting at one knot with a bearing of 226 degrees.
I am on fire and in need of immediate assistance.
There are 35 persons on board with a couple of persons severely burned.
My vessel is a car-carrier, 200 meters long and 32 meters wide, and has a black color hull with red trim.
This is the Maritime Dream, over.

語彙

1. send a mayday：メーデー［遭難信号］を発信する
2. drift：〔動詞〕漂う，漂流する

3. bearing：〔名詞〕方位，方角
4. in need of：〜を必要としている
5. immediate assistance：緊急の援助
6. severely burned：重度の火傷を負った
7. hull：〔名詞〕船体
8. trim：〔名詞〕装飾，線

フォーカスする文法／表現

【英語ワザ-1】 There are構文／ Here is 構文
◆ There is 単数名詞．　There are 複数名詞．「〜があります」
◆ Here is 単数名詞．　Here are 複数名詞．「ここにあります」

- There is/are構文／ Here is/are構文：「〜があります」「ここに〜がいます」という構文は，初めて話題にのぼるものを紹介する。たとえば，「むかしむかし，あるところに〜がいました」というように初めて登場させるものを紹介する場合に，There are構文やHere is構文を使う。
- 「はい，どうぞ」と提出して見せるときにはHere it is.やHere they are.と言う。

練習A　　次の英文を訳しなさい。

① There **are** a number of key documents [that you always need to carry with you].

訳 _____

② There **is** a vessel on the same course, right astern, distance 1.5 miles.

訳 _____

③ There **are** 35 persons on board with a couple of persons severely burned.

Story 15
Sending a Mayday Signal

訳 _____

④ **Is** there any leakage or abnormal sound？ No, sir.

訳 _____

⑤ Here **are** the documents for customs and immigration．

訳 _____

⑥ Do you have a crew list？ Yes, here it is.

訳 _____

⑦ Do you have passports？ Yes, here they are.

訳 _____

【英語ワザ-2】動名詞 (Ving) で指示の復唱
◆動詞＋ing は動名詞として知られているが，船上のコミュニケーションでは，情報伝達を確実に行うために，指示の復唱にこの形が使われている。
(例1) G：The captain has just ordered to send a mayday.
　　　A：Roger, sir. Sending a mayday signal. Switching to VHS channel 16.
　　　＊mayday：(船舶の無線電話による) 遭難信号
　　　＊Roger：(無線) 了解
(例2) F：Close the starting air intermediate valve, and blow down the starting air line.
　　　B：Closing the starting air intermediate valve, and blowing down the starting air line.

＊starting air：始動空気
＊intermediate valve：中間弁
＊blow down：ボイラ水を排出する

（例3）G：Heave up port cable.

A：Heaving up port cable.

＊heave up：索で引き揚げる
＊port cable：左舷錨鎖

練習B　作業を安全・確実に遂行するために，指示を動名詞を使って声に出して復唱してから，書きなさい。

① F：First, open the strainer bypass valve, and close the sea water inlet and outlet valves.

　　B：_____

② F：Relieve the pressure inside the housing and dismantle the cover.

　　B：_____

③ G：Change to VHF channel 22 (two-two).

　　A：_____

④ G：Stand by on VHF channel 16 (one-six).

　　A：_____

【英語ワザ-3】 高さ，幅，深さなどの表現

◆200 meters long ／ the length is 200 meters

　32 meters wide ／ the width is 32 meters

(例) The ship is 200 meters long. = The length of the ship is 200 meters.
　　　The ship is 32 meters wide. = The width of the ship is 32 meters.

練習C　　次の単語(形容詞)の名詞形を書きなさい。

① long ⇒ _____

② wide ⇒ _____

③ deep ⇒ _____

④ high ⇒ _____

練習D　　下記の船舶Aに関する情報を基に, (　　　)に適切な語句を書きなさい。

① How long is the ship?

　　The (　　　　) of the ship is (　　　　).

② How wide is the ship?

　　The (　　　　) of the ship is (　　　　).

③ How deep is the ship?

　　The (　　　　) of the ship is (　　　　).

＜船舶Aに関する情報＞
　　全長：53.59m　垂線間長：46.00m　幅：10.00m
　　深さ(上甲板／第二甲板)：5.40／3.40m　喫水：3.20m
　　国際総トン数：731トン　登録総トン数：231トン　燃料タンク容積：56m^3

清水タンク容積：74m³

練習E　下記の船舶Bに関する情報を基に，（　　　）に適切な語句を書きなさい。

① What is the length of the ship?

　This ship is (　　　　　　　).

② What is the width of the ship?

　This ship is (　　　　　　　).

＜船舶Bに関する情報＞
　全長：241m　垂線間長：46.00m　幅：29.6m　喫水：7.8m
　国際総トン数：50,142トン　最大速力：21ノット

◇リーディング・海技試験英語対策

次の英文を日本文になおしなさい。

Distress traffic includes all messages relating to immediate assistance required by persons, aircraft, or marine craft in distress, including medical assistance. Distress traffic may also include SAR communications and on-scene communications. Distress calls take absolute priority over all other transmissions; anyone receiving a distress call must immediately cease any transmissions which may interfere with the call and listen on the frequency used for the call.

　　(2007年版IAMSAR MANUAL Volume IIより　http://www.bsmrcc.com/files/legal2.pdf)

《語　注》

1. distress：〔名詞〕遭難，災難，救難
2. traffic：（情報などの）やりとり，通信
3. distress traffic：遭難通信
4. immediate assistance：緊急の援助
5. SAR (Search And Rescue) communication：捜索救難通信
6. on-scene communication：現場通信
7. absolute：〔形容詞〕絶対的な，完全な
8. take priority over ～：～よりも優先権を得る
9. transimission：〔名詞〕送信
10. interfere：〔動詞〕妨害する
11. frequency：〔名詞〕周波数

本の紹介

『海技士1N 徹底攻略問題集』『海技士2N 徹底攻略問題集』
東京海洋大学海技試験研究会編
海文堂出版

海運業界はさまざまな国の人たちとのやりとりが基盤にある，もっともグローバルな業界であるといっても過言ではありません。そのような世界を舞台とする企業はその母国の法律だけでなく国際的な条約や取り決めをもとにビジネスを行わなくてはなりません。『海技士1N／2N 徹底攻略問題集』の英語問題は，このような海事の国際条約を正しく理解し，日本語に訳す練習を多く掲載しています。条約などの英語は長くて最初は読解に苦労するかもしれません。しかし慣れてくると同じようなパターンがよくでてくることに気づくでしょう。この問題集を通して繰り返し勉強をし，単に試験合格を目指すだけでなく，国際条約をきちんと読解できる力を未来の海運業界の担い手としてつけていきましょう。

Story 16　We Are Ready for Unmanned Operation

Kaiくんの乗船する船は，機関室の当直なしで主機を24時間運転できる設備を備えたMゼロ船です。日中は機械類の点検や整備が行われていますが，夜間は無人化運転が行われます。この無人化運転に備えて，毎日午後3時半から5時まで，Mゼロチェックを行います。今日の点検は先輩の三等機関士とKaiくんの2人で行います。まだ慣れないKaiくんはマニュアルにあるチェックリストを見ながらの作業です。

◎海しごと＆海事知識

Mゼロ船：機関士の当直なしで，主機を24時間運転できる設備を備えた船をMゼロ (Machinery Space Zero Person) 船と呼ぶ。船橋からの主機の遠隔コントロールが可能で，かつ機関室の異常を知らせる警報装置，異常時に自動的に主機を減速する装置など，無人運転のための多くの規定を満たした船を意味する。Mゼロ船といっても，常時，機関室無人化運転が行われているわけではなく，通常航海中は日中，機械類の点検や整備を行い，夜間無人化運転を行う。また，船舶の通行量の多い海域や，天候が悪化したときなどは，機関長の判断で機関室当直が行われる。現在では外航船や内航大型船のほとんどがMゼロ船となっている。　　　　　　　　　　　　　　　　　　　　　　（『船しごと、海しごと。』より）

◎ダイアログ

3/E：OK, next, let's check tank levels.　How are the main and auxiliary engine header tanks?

Kai：They are full, sir.

3/E：Good.　How about the stern tube header tank?

Kai：Full, sir.

3/E：What about the waste oil tank level and the temperature?

Kai：The waste oil tank is full, and the temperature is normal, sir.

3/E：Good.　Let's check the main engine next.　Do you see any abnormalities

Story 16
We Are Ready for Unmanned Operation

on the main engine cylinder tops?

Kai : Abnormalities?

3/E : Is there any leakage or movement? Or do you detect any temperature or pressure outside the normal parameters?

Kai : No, everything is within the normal parameters, sir.

3/E : Lastly, ensure that all the fire detection sensors are switched on.

Kai : All the sensors are switched on, sir.

3/E : Good. Notify the bridge that we are ready for unmanned operation.

Kai : Yes, sir.

語彙

1. auxiliary engine：補助エンジン
2. header tanks：ヘッダタンク，配水タンク
3. stern tube：船尾管
4. waste oil tank level：廃油タンクレベル
5. abnormality：〔名詞〕異常
6. parameter：〔名詞〕パラメータ，変数(値)
7. fire detection：火災探知，火災検知
8. be switched on：スイッチが入っている
9. notify：〔動詞〕報告する，通知する
10. unmanned operation：無人運転，無人操作

フォーカスする文法／表現

【英語ワザ-1】 前置詞（位置関係を表す）：前置詞句が修飾する語は？

◆位置関係を表す前置詞
- 位置関係を表す前置詞にはat, in, on, over, against, through, outside, insideがある。
- 前置詞の基本的なイメージをおさえる。

◆前置詞句が修飾する単語を探せ！⇒ 前置詞句はどこにかかるか？
- 前置詞句（前置詞＋名詞）は，前置詞句より前にある「名詞」あるいは「動詞」を修飾する。
- 練習Aの①～③では，前置詞句を下線で示し，それが修飾する単語を☐で示してある。修飾される単語がどれかを意識すると，どこからどう訳すのかわかりやすい。

練習A 　以下の英文を和訳しなさい。また，④～⑥の英文では，下線部（前置詞句）がどの単語を修飾しているか，矢印で示しなさい。

① The document shows that you have received [vaccination] against yellow fever.
- against：「向かい合う力」のイメージから，「～に対して，～に反対して，～に備えて」を表す。

訳 _____

② A muster list [is posted] throughout the ship, including at the bridge, engine room and crew accommodation areas.
- through：トンネルを抜けていくイメージから「～を通って，～を通して」の意味。
- throughout：トンネルの「始めから終わりまで」というイメージ。「～中いたるところに」の意味。

訳 _____

Story 16
We Are Ready for Unmanned Operation

③ [Speed] through the water is 11.5 knots.

訳 _____

④ With this shore pass, a crew will be given access <u>through gates</u> only for the area [that the vessel is berthed].
- with：「〜と一緒に」と，さまざまなものとのつながりを表す。
「this shore passと一緒に」という意味なので，「〜を持っていると」という意味になる。

訳 _____

⑤ Do you detect any temperature or pressure <u>outside the normal parameters</u>?
- in：「容器」のなかに ⇔ out：「容器」の外に
- inside：〜の内部に（へ），〜の内側に
- outside：〜の範囲を超えて，〜以外に，〜の外側に（で）

＊detect：見つける，認める，〜の存在を発見する
＊parameters：パラメータ，変数（値）

訳 _____

⑥ Then, relieve the pressure <u>inside the housing</u> and dismantle the cover.

訳 _____

【英語ワザ-2】語彙力アップ：接頭語＆接尾語
◆接頭語に注目すると，イメージが湧き，単語の意味が推測できる。
ab-は「離れた」の意味を持つ。
- Abuse：ab-「離れた」＋use「〔動詞〕使う，〔名詞〕使い方」
 推測／イメージ：普通からかけ離れた使い方

単語の意味：〔名詞〕乱用，誤用，虐待
〔動詞〕乱用する，誤用する，虐待する
- Abnormal：ab-「離れた」+ norm「規格」+ -al（名詞，形容詞をつくる接尾語）
推測／イメージ：規格から離れている
単語の意味：〔形容詞〕異常な

練習B　与えられた接頭語とそれに続く単語の意味を参考に，単語の意味を推測し，その後，辞書で単語の意味を調べなさい。

① Abstain：ab-「離れた」+ stain「保つ，含む」

　推測／イメージ：_____

　単語の意味：_____

② Abduct：ab-「離れた」+ duct「導く」

　推測／イメージ：_____

　単語の意味：_____

◆接尾語に注目すると，単語の品詞が推測できる。
- -ageは名詞をつくる接尾語で，結果や状態を表す。
  leakage：leak〔動詞〕「漏れる」+ -age =〔名詞〕漏れ，漏出
- -ityは名詞をつくる接尾語で，性質，状況，程度などを表す。
  abnormality：abnormal〔形容詞〕「異常な」+ -ity =〔名詞〕異常
- -mentは動詞から名詞をつくる接尾語で，結果，状態，動作，手段を表す。
  measurement：measure〔動詞〕「測る」+ -ment =〔名詞〕計測

練習C　与えられた単語の意味を参考にして，接頭語がついたときの単語の意味を空欄に書きなさい。

① shortage：short〔形容詞〕「不足している」+ -age = _____

② package：pack〔動詞〕「詰める，包む」+ -age = _____

③ usage：use〔動詞〕「使う，使用する」+ -age = _____

④ development：develop〔動詞〕「発展させる，開発する」+ -ment =

_____

⑤ argument：argue〔動詞〕「議論する」+ -ment = _____

⑥ shipment：ship〔動詞〕「輸送する」+ -ment = _____

【英語ワザ-3】 動詞の種類(他動詞／自動詞)
◆英語では，動詞の働きを他動詞と自動詞の2種類に分類する。
◆辞書では，他動詞を*vt*，自動詞を*vi*という記号で示している。
◆自動詞と他動詞の両方の働きをする動詞もある。
 (例) run (*vi*)：〜が走る
　　　run (*vt*)：〜を運営する，〜を実行する
＜他動詞＞
 ・他動詞は，その右側に目的語を必要とする。目的語の働きをするのは名詞，代名詞，動名詞(Ving)，不定詞(to V)である。
 (例1) I take the bus.　私はバスに乗る。
 (例2) I run a store.　私は店を経営しています。
 ・動詞の意味が「〜をどうする」「〜にどうする」タイプならば，他動詞が使われる。
 ・対象物のある作業に関する文は「〜をどうする」ということを表すので他動詞が使われる。

<自動詞>
- 自動詞は，目的語が不要。主語と自動詞だけで，文として完結する。
  （例）I ran.　私は走った。
- 自動詞の後ろに，さらに説明を加えたい場合はto, at, for, inなどの前置詞が必要になる。
  （例）I ran [in the morning].
- 自動詞の後にくる副詞／副詞句は動詞を修飾する。
  （例）He runs very fast.
- 他動詞か自動詞かは，動詞の右側に名詞がきているかどうかで判断できる。
  ① My father **walks** our dog early in the morning.
    walkの後にour dog〔名詞〕が付いているのでwalkは他動詞「〜を散歩させる」。
  ② My father **walks** very fast.
    walkの後にvery fast〔副詞〕が付いているのでwalkは自動詞「〜が歩く」。
  ③ I **got** up at 6：30 yesterday.
    gotの後にup at 6：30〔副詞＋前置詞句〕が付いているのでgotは自動詞。
  ④ I **got** a wonderful present from my family on my birthday.
    gotの後にa wonderful present〔名詞〕が付いているのでgotは他動詞「〜を手にする」。
    from my familyとon my birthdayは前置詞句（前置詞＋名詞）と呼ばれ，gotを修飾している。

練習D　　下線の動詞の種類（他動詞か自動詞か）を右の［　　］内に書きなさい。

① So, does it snow over there in winter?　　　　　　［　　　］

② It is raining very hard.　　　　　　　　　　　　　［　　　］

③ You have to explain to them what it is.　　　　　　［　　　］

④ People can pick the cooked ingredients that they like from the pot.
[   ]

⑤ How do I eat it? [   ]

⑥ Let's check the main engine next. [   ]

⑦ Do you see any abnormalities on the main engine cylinder tops?
[   ]

⑧ Do you detect any temperature or pressure outside the normal parameters?
[   ]

⑨ Ensure that all the fire detection sensors are switched on. [   ]

⑩ Notify the bridge that we are ready for unmanned operation. [   ]

◇リーディング・海技試験英語対策

次の英文を日本文になおしなさい。

Performing the Engineering Watch

When the machinery spaces are in the manned condition, the officer in charge of the engineering watch shall at all times be readily capable of operating the propulsion equipment in response to the needs for changes in direction or speed.

When the machinery spaces are in the periodic unmanned condition, the designated duty officer in charge of the engineering watch shall be immediately available and on call to attend the machinery spaces.

(STCW条約より)

《語　注》

1. engineering watch：機関当直
2. machinery space：機関室
3. manned：〔形容詞〕有人の
4. readily：〔副詞〕すぐに，ただちに
5. capable：〔形容詞〕〜ができる，〜する能力がある
6. propulsion equipment：推進装置
7. in response to：〜を受けて，〜に応じて
8. periodic：〔形容詞〕定期的な
9. designated：〔形容詞〕指定された，指名された
10. duty officer：当直士官
11. immediately：〔形容詞〕直ちに
12. on call：待機して
13. attend：見張る，用心する，（職務として）世話をする

=============== 本 の 紹 介 ===============

『船しごと、海しごと。』
商船高専キャリア教育研究会編
海文堂出版

働くということを考えるところから，海運業界におけるさまざまな職種の紹介や業界分析，そして就職活動に関する情報までをカバーする，海や船を舞台とする仕事に興味のある学生にとって最適な入門書です。現在，海運業界で活躍する先輩からのメッセージも多く掲載し，海運業界での仕事の様子がいきいきと伝わってきます。

Story 17　Welcome Aboard

荷降ろしをするスペインの港に入港しました。ここでは検疫と税関および船舶代理店のエージェントが乗船してきました。Kouくんは，検疫と税関および乗組員の上陸の手続きなどのやりとりをまかせられました。先輩サードオフィサーから教えてもらったことや，すべて英語で書かれたマニュアルを思い出しながら，どうにかやりとりをしています。

◎海しごと＆海事知識

船舶の入出港時には以下のようなさまざまな手続きが必要になります。これら手続きの多くは船舶代理店が船長の代行として行っています。

□入港前に必要な手続き
- 入港通報　　　　　　　　　→ 入国管理局，検疫所
- 船舶保安情報　　　　　　　→ 海上保安部署
- 係留施設使用許可申請　　　→ 港湾管理者
- 危険物荷役許可申請など　　→ 港長
- 航路通報　　　　　　　　　→ 海上交通センター
- 事前通報　　　　　　　　　→ 信号所
- 保障契約情報　　　　　　　→ 地方運輸局

□入港時に必要な手続き
- 入港届　　　　　　　　　　→ 税関,入国管理局,検疫所,港長,港湾管理者
- 明告書　　　　　　　　　　→ 検疫所
- 乗組員上陸許可申請　　　　→ 入国管理局

□出港時に必要な手続き
- 出港届　　　　　　　　　　→ 税関，入国管理局，港長，港湾管理者
- とん税および特別とん税納付申告 → 税関

（『船しごと、海しごと。』より）

◉ダイアログ

Kou　　：Are you Mr. Wales of the Global Shipping Agency?
Agent：Yes.
Kou　　：My name is Kou, a junior third officer of this ship. Welcome aboard. Here are the documents for customs and immigration.
Agent：Do you have a crew list?
Kou　　：Yes, here it is.
Agent：Passports?
Kou　　：Yes, here they are. Do we keep the passports? Or do you keep them?
Agent：We do.
Kou　　：OK. When do we get shore passes?
Agent：They are ready. As you can see, crew shore passes at this harbor are swipe cards. With this, crew will be given access through gates only for the area that the vessel is berthed.

Kou　　：OK.
Agent：Crew shore passes must be visible at all times while the crew are within the port boundary. Make sure that all crew keep to the designated walkways within the port boundary.
Kou　　：OK, I will relay the message to the crew.
Agent：Sign this letter of receipt if you acknowledge the shore access procedures.

語彙

1. shipping agency：船舶代理店
2. customs and immigration：通関および出入国管理
3. swipe card：磁気カード
4. berth：〔動詞〕停泊する，停泊させる
5. boundary：〔名詞〕境界
6. designated：〔形容詞〕指定された
7. walkway：〔名詞〕歩道，通路
8. relay the message：伝言を伝える
9. acknowledge：〔動詞〕同意する，承認する
10. procedure：〔名詞〕手順，手続き

フォーカスする文法／表現

【英語ワザ-1】海仕事関連語句や表現
◆乗船，下船手続きなどに関連する単語を覚える
① customs：（複数形で）関税，税関，通関手続き
　（例1）We must pay customs on goods (that) we have bought.
　（例2）Here are the documents for customs and immigration.
　（例3）As a shipping agency, we prepare documents for the customs and harbor services.
② immigration：（空港，港などでの）出入国管理，入国審査
③ passport：旅券，パスポート，通航証
　（例）A passport is the most important identification document while (a person is) overseas.
④ shore pass：寄港地上陸許可証 (SP)
　（例1）Crew shore passes must be visible at all times while crew are within the port boundary.
　（例2）When can we get shore passes for all the crew?

⑤ quarantine：〔名詞〕検疫(所)，〔動詞〕検疫する
　（例）The Departmernt of Agriculture officers quarantined the ship.

練習A　以下の文が説明する単語として最もふさわしいものを下の【　】内から選びなさい。

① An official document issued by the government of a country that identifies someone as a citizen of that country and that is usually necessary when entering or leaving a country.

　　　　　　　　　　　【解】_____

② Taxes or fees that are paid to the government when goods come into or go out of a country.

　　　　　　　　　　　【解】_____

③ The place at an airport or country's border where government officials check the documents of people entering that country.

　　　　　　　　　　　【解】_____

④ A permit issued by the immigration authorities for the sailors to spend their time on land.

　　　　　　　　　　　【解】_____

⑤ A system of measures maintained by governmental authority at ports, etc., for preventing the spread of disease.

　　　　　　　　　　　【解】_____

【shore pass,　customs,　passport,　immigration,　quarantine】

Story 17 Welcome Aboard

【英語ワザ-2】海仕事関連語句や表現

◆「船に乗る」に関連する単語を，用例で覚える

① board：〔他動詞〕人が（飛行機，船，列車）に乗り込む
　　（例1）You will be boarding our ship on your first job assignment.
　　（例2）You will board the ship at Nagoya port.

② aboard：〔副詞〕〔前置詞〕（飛行機，船，列車）に乗って
　　（例1）Welcome aboard!
　　　　　　ご搭乗（乗船，乗車）ありがとうございます。
　　（例2）There are 30 crewmembers aboard this ship.
　　　　　　この船には30名の乗組員が乗船している。

③ on board：〔副詞〕〔前置詞〕乗り物（飛行機，船，列車）に乗って＝aboard
　　（例1）The third officer will train you on the daily responsibilities and tasks on board the ship.
　　（例2）work on board a ship：乗船中の仕事
　　（例3）persons on board：乗船中の人々

④ get on：〔自動詞〕人が（バス，馬など）に乗る
　　（例1）When the bus stopped, we got on.
　　（例2）That is the ship (which) you are getting on!

⑤ get off：〔自動詞〕
　　人が（乗り物，馬など）から下りる，離陸する，旅に出発する
　　（例）He got off the plane in New York.

⑥ embark：〔自動詞／他動詞〕～が乗船する，～を船に乗せる
　　（例1）We embarked on the ship in January.
　　（例2）embarkation procedure：出国手続き

⑦ disembark：〔自動詞／他動詞〕～が下船する，（乗客を）船から降ろす
　　（例1）This is the port where we are going to disembark.
　　（例2）disembarkation card：入国（記録）カード

練習B 以下の日本文を【　】内の単語を用いて英文にしなさい。

① 私はスペインからこの船に乗船しました。【board】

解 _____

② 彼は次の港で下船します。【get off】

解 He _____

③ その船が衝突し，乗組員10名が怪我をした。
　【collided, crew members, on board, were injured】

解 The ship _____

④ 船はすべての乗組員をKosen港で降ろした。【disembark】

解 The ship _____

【英語ワザ-3】海仕事関連語句や表現

◆海仕事で頻繁に使用される動詞を，用例で覚える

① check：〔他動詞〕（人，物，事）を（よい状態か，正しいか）検査する，
　　　　　　　　　調べる
　　　　　　〔自動詞〕調べる，調査する
　（例1）Let's <u>check</u> tank levels.
　（例2）Don't forget to <u>check</u> for any leakage from the strainer cover.

② provide：〔他動詞〕～を用意する，提供する
　（例1）The third officer, as the safety officer on the ship, must <u>provide</u> an orientation on emergency procedures as well as vessel safety and survival equipments.
　（例2）The housing is <u>provided</u> with a spindle guide.

③ make sure：〔他動詞〕～を確かめる，(必ず～するようにthat以下)を手配する，注意する
   (例1) <u>Make sure</u> (that) you understand the captain's standing order.
   (例2) I must <u>make sure</u> that there are no suspicious persons or unidentified vessels in the port area.
④ ensure：〔他動詞〕
   ～を確実にする，保証する，確保する，必ず手に入るようにする
   (例1) We <u>ensure</u> a berth for the incoming ship, and arrange for the pilot and tugs if necessary.
   (例2) Lastly, <u>ensure</u> that all the fire detection sensors are switched on.
⑤ relay A to B：A(伝言，ニュースなど)をBへ伝える
   (例1) I also <u>relay</u> important information such as weather, or traffic condition in the harbor to the ship.
   (例2) I will <u>relay</u> the message to the crew.
⑥ be ready for 名詞／be ready to V：
   すぐに(いつでも)～できる(用意ができている)
   (例1) We <u>are ready for</u> unmanned operation.
   (例2) <u>Ready for</u> main engine trial.
⑦ be familiar with 名詞：～を熟知している，よく知っている，～に通じている
   (例1) All the crew are <u>familiar with</u> how to use each emergency equipment.
   (例2) Are you <u>familiar with</u> the job of an agent, like me?

練習C　以下の日本文を【　】の単語をすべて用いて英文にしなさい。
① すべての乗組員は必ず港内の指定された通路内を歩くようにしてください。
【all crew, the port boundary, the designated walkways, make sure, walk on, that, within】

_____

② 次に，主エンジンを検査しましょう。【the main engine, check, let's, next】

解 _____

③ すべての乗組員は非常時の手順に精通していなくてはならない。
【must, emergency procedures, be familiar with, all crew】

解 _____

④ 錨を降ろす準備ができています。【let go, are ready, anchor, we, to】

解 _____

⑤ このメッセージを船長に伝えてください。
【this message, relay, please, the captain, to】

解 _____

⑥ 入港してくる船の係留場所を確認しなさい。
【a berth, ensure, the incoming ship, for】

解 _____

◇リーディング・海技試験英語対策

以下の英文を日本語に訳しなさい。

Customs, Department of Agriculture and Immigration clearance must be completed prior to going ashore.

・Please stay on board. No persons other than a Department of Agriculture

or Customs officer is allowed to board your craft, nor can any person, animal or article leave the craft until you have been given full clearance;
- Depending on your arrival time, Customs and Department of Agriculture may require all persons to remain on board overnight before clearing you the following day;
- Don't throw any waste or foodstuffs overboard while you're in Australian waters or while you are moored. Use designated biosecurity disposal points;
- Keep all food and animals secure until your vessel has been inspected by Department of Agriculture officers;
- Don't trade foodstuffs with other overseas vessels;
- Keep your vessel free of insects.

(オーストラリア政府税関国境警備局ホームページ
http://www.customs.gov.au/site/page4360.aspより)

《語 注》

1. Department of Agriculture：
   農務省
2. clearance：〔名詞〕
   許可，通関手続き
3. be allowed to：
   〜することを許可されている
4. craft：〔名詞〕船，飛行機
5. article：〔名詞〕品物
6. require：〔動詞〕〜を必要とする，
   命じる，要求する
7. remain：〔動詞〕とどまる，残る
8. clear：〔動詞〕出入港手続きを
   完了する，通関する
9. foodstuffs：〔名詞〕食料品
10. overboard：〔副詞〕船外に
11. moored：〔形容詞〕係留してある，
    停泊している
12. biosecurity：バイオセキュリティ
    （国内の家禽に対する保健上の
    予防措置）
13. disposal：〔名詞〕処分，廃棄
14. secure：〔動詞〕安全に保管する，
    外に出さない
15. inspect：〔動詞〕〜を検査する，
    点検する
16. trade：〔動詞〕取り引きする，
    交換する，売買する
17. insect：〔名詞〕虫

Story 18　I Have Enclosed Our Damage List

長い航海もあっというまに最後の港となりました。すべての荷降ろしが完了し，サーベイヤーによる検査を受け，必要な修理がすべて行われました。同サーベイヤーより母港であるKosen港に向けた耐航性証明書が発行され，いよいよ，日本へ向けて航海です。船の上では，Kouくんが過去に書かれた報告書を参考に，修理発注および終了の報告書を英語で作成しています。

◎海しごと＆海事知識

すべての船舶はその乗組員や耐航性など船舶に関する各種証明書をつねに提示できる状態にしておかなくてはなりません。これら証明書は船舶の旗国によって発行され，各国の港湾管理官により検査が行われます。
以下に挙げたのは，船舶に常時保管されていなくてはならない証明書類の一部です。

1. Certificate of Nationality：船舶国籍証書
2. International Tonnage Certificate：国際トン数証書
3. International Load Line Certificate：国際満載喫水線証書
4. Intact stability booklet：非損傷時復原性手引書
5. Certificates for masters, officers or ratings：
   船長，士官，部員の資格証明書，海技免状など
6. International Oil Pollution Prevention Certificate / Oil Record Book：
   国際油汚染防止証書
7. Shipboard Oil Pollution Emergency Plan：油濁防止緊急措置手引書
8. Garbage Management Plan：廃棄物処理計画
9. Garbage Record Book：廃棄物記録簿
10. Cargo Securing Manual：貨物固定マニュアル
11. Safety Management Certification (SMC)：安全管理証書

（参照：http://wwwtb.mlit.go.jp/chubu/shizuoka/shimizu/

Story 18
I Have Enclosed Our Damage List

kensayougo.htm#%81yAFS%81zInternational%20A
http://www.imo.org/OurWork/Facilitation/FormsCertificates/Pages/Default.aspx)

◉ダイアログ

We hereby report that based on the surveyor's recommendation, necessary repairs were ordered and completed, and that the ship was granted the seaworthy certificate for our homebound voyage.

We arrived here in Spain at 10:00 a.m. on the 5th of August and after discharging the entire cargo, we underwent a survey by Mr. Park, of Global Marine Surveyors Ltd. In accordance with his recommendations, we ordered the necessary repairs by Pacific Dockyard Co. Ltd. Upon completion of the repairs on the 8th of August, Mr. Park granted the seaworthy certificate for our homebound voyage to Kosen port, Japan.

I have enclosed our damage list, a sketch of the damaged parts, and copies of each survey report.

The bill of repairs has not reached us yet, however, we will be in touch with you as soon as we receive it.

If you have any questions or concerns regarding this matter, please feel free to contact me.

語彙

1. hereby：〔副詞〕これによって
2. surveyor：〔名詞〕サーベイヤー，鑑定人
3. in accordance with：〜に従って

4. grant：〔動詞〕与える，発行する
5. seaworthy certificate：耐航性証明書
6. homebound voyage：帰航
7. discharge：〔動詞〕（人や荷）を降ろす，荷下ろしする
8. entire：〔形容詞〕すべての
9. cargo：〔名詞〕貨物
10. underwent：〔動詞〕undergoの過去形
11. undergo：〔動詞〕〜を受ける，〜を経験する
12. survey：〔動詞〕調査する
13. dockyard：〔名詞〕造船所
14. complete：〔動詞〕完了する，終える
15. completion：〔名詞〕完成，終了，完了
16. enclose：〔動詞〕〜を同封する
17. bill：〔名詞〕請求書

【英語ワザ-1】 海仕事関連語句や表現：動詞
◆動詞をしっかりおさえて英語らしい文章を！
英語の文章では，動詞が中心的な役割を果たすことが多い。英語の文章の作成には，まずは「誰が／何が」＝主語を意識し，その次に，その主語の動作や行動を表す動詞は何か考えるとよい。
同じ主語などが続く場合は，文を受け身形にすると文章全体が自然になる。
（例）We ordered necessary repairs. ⇒ Necessary repairs were ordered.
◆海仕事で頻繁に使用される動詞を，用例で覚える。その際，反意語や同意語も意識して覚えるとよい。
① repair：修理する
② order：注文する
③ complete/finish：完了する，終了する
④ arrive：到着する ⇔ depart：出発する
⑤ discharge：（荷・人を）降ろす，unload：（荷を）降ろす ⇔ load：荷積みする
⑥ undergo：〜を受ける，〜を経験する

Story 18
I Have Enclosed Our Damage List

⑦ grant：与える，発行する
　be granted 〜：〜を与えられる
⑧ survey：調査する
⑨ enclose：〜を同封する

練習A　前掲のダイアログの動詞を□で囲み，その動詞の主語に下線を引きなさい。

練習B　日本文の意味に合うように以下の空欄に最も適切な動詞を書きなさい。

① 主エンジンの確認作業を完了しました。

　We have (　　　　　) checking the main engine.

② 今日は冷却水タンクの修理をします。

　I will (　　　　　) the cooling water tank today.

③ (この手紙に) 私の子どもの写真を同封しました。

　I have (　　　　　) a picture of my children with this letter.

④ 必要な部品および道具を直ちに注文しなくてはならない。

　Necessary parts and tools need to be (　　　　　) immediately.

⑤ 怪我をした乗組員は，次の港で下船し，手術を受ける予定だ。

　The injured crewmember will (　　　　　) at the next port and
　(　　　　　) a surgery there.

171

⑥ 次の港で乗組員全員を降ろします。

　　The ship is going to (　　　　　) all the crew members at the next port.

⑦ 港の管理区域に入る許可をもらった。

　　We were (　　　　　) a permission to access the controlled area of the harbor.

【英語ワザ-2】 語彙力アップ：品詞

◆新しい単語を学んだら，その単語の他の品詞も調べて一緒に覚えて語彙力をアップ
- 品詞が変わることで語尾が変わるものもあれば，品詞が変わっても単語が変わらないものもある
  - （例1）complete〔動詞〕⇒ completion〔名詞〕
  - （例2）complete〔動詞〕⇒ complete〔形容詞〕
  - （例3）bill〔名詞〕⇒ bill〔動詞〕
- 海仕事で頻繁に使われる単語のさまざまな品詞を用例で覚える
  - （例）The <u>bill</u> of repairs has not reached us yet.
    ＝ They (Pacific Dockyard Co. Ltd) haven not <u>billed</u> us yet.

練習C　以下の単語を（　　）の品詞にしなさい。またその意味を辞書で調べて書きなさい。

① arrive〔動詞〕

　　（名詞）【　　　　　　　　】　意味：

　　（形容詞）【　　　　　　　　】　意味：

② seaworthy〔形容詞〕

　　（名詞）【　　　　　　　　】　意味：

③ report〔動詞〕

　　（名詞）　【　　　　　　　　　】　意味：

　　（形容詞）【　　　　　　　　　】　意味：

④ recommendation〔名詞〕

　　（動詞）　【　　　　　　　　　】　意味：

　　（形容詞）【　　　　　　　　　】　意味：

⑤ necessary〔形容詞〕

　　（名詞）　【　　　　　　　　　】　意味：

　　（動詞）　【　　　　　　　　　】　意味：

　　（副詞）　【　　　　　　　　　】　意味：

⑥ receive〔動詞〕

　　（名詞）　【　　　　　　　　　】　意味：

　　（名詞）　【　　　　　　　　　】　意味：受け取る人

練習D　下線部[that 以下]のなかの主語と述語動詞を抜き出しなさい。

① We **have to make sure** [that goods are delivered in a timely manner].

　　主語S ＿＿＿＿＿＿＿＿＿＿＿＿　述語動詞V ＿＿＿＿＿＿＿＿＿＿＿＿

② It **is** very likely [that the strainer is clogged].

　　主語S ＿＿＿＿＿＿＿＿＿＿＿＿　述語動詞V ＿＿＿＿＿＿＿＿＿＿＿＿

③ I **must check** [that there are no suspicious persons or unidentified vessels in the port area].

　　主語S _____　　述語動詞V _____

④ **Ensure** [that all the fire detection sensors are switched on].

　　主語S _____　　述語動詞V _____

⑤ **Notify** the Bridge [that we are ready for unmanned operation].

　　主語S _____　　述語動詞V _____

⑥ **Make sure** [that all crew keep to designated walkways within the port boundary].

　　主語S _____　　述語動詞V _____

⑦ We hereby **report** [that the ship was granted the seaworthy certificate for our homebound voyage].

　　主語S _____　　述語動詞V _____

⑧ **Ensure** [that you must wear an appropriate eye protection in all designated area].

　　主語S _____　　述語動詞V _____

⑨ We **must understand** [that safety rules vary depending on the type and size of the plant].

　　主語S _____　　述語動詞V _____

Story 18
I Have Enclosed Our Damage List

⑩ We **learned** [that generators change mechanical energy into electrical energy].

　　主語S _____　　述語動詞V _____

【英語ワザ-3】 英文で書く手紙や報告書のフォーマットや表現
英語の文章の特徴として，①導入，②具体例，③まとめ，という3部構成が多い。
導入でその文章の主題／主なポイントを先に明確に述べ，それに続く段落で具体例や主題文を支えるための説明がなされる。
まとめでは，具体例で述べたことの要約や，次のステップやアクションなど，未来志向の文が多い。

練習E　　Story18のダイアログを①導入，②具体例，③まとめに分類すると，どのように区切るのが最も適切か，p.169の英文中にそれぞれを【　】で囲みなさい。

◆手紙やメールなどでの導入そしてまとめで定型文がよく使われる
＜導入＞　動詞を中心に文書の目的を明確に述べる。that以下には主語＋動詞の文が置かれる。
　報告：We hereby <u>report</u> that 主語＋動詞
　　　　 We wish to <u>report</u> that 主語＋動詞
　通告／通知：We hereby <u>notify</u>/<u>inform</u> you that 主語＋動詞
　確認：We hereby <u>confirm</u> that 主語＋動詞
　警告：We are writing this e-mail to <u>warn</u> you that 主語＋動詞
　問い合わせ：We are writing this letter to <u>inquire</u> about 名詞
＜具体例＞　具体例としては，いろいろなパターンがある。
　・起こったことを時系列に説明（過去形）
　・理由
　・手順など
＜まとめ＞
　|同封／添付|
　・I have enclosed our damage list.

- Enclosed is the sketch of damaged parts.
- I have attached a bill for the repair.
- Attached to this letter is a bill of the repair.
- Please find the enclosed bill of the repair.
- Please see the attached report for your reference.

質問／問い合わせ

- If you have any questions or concerns regarding this matter, please feel free to contact us.
- Should you have any questions or concerns, do not hesitate to contact us.

お願い

- Please let us know when the repairs will be completed.
- Please reply to us by the end of this week with the quote for the repair.

練習F 以下のメール／手紙の導入文を英語に訳しなさい。

① ご注文いただきました修理が完了したことを報告いたします。

訳 _____

② このメールをもちまして，Kosen港を出港したことを報告いたします。

訳 _____

③ このメールをもちまして，到着が1日遅れる (be delayed) ことを通知いたします。

訳 _____

④ 新しい商品 (a new product) についての情報を問い合わせたく，このメールを書いております。

訳 I am writing_____

Story 18
I Have Enclosed Our Damage List

練習G　練習Fの導入文②〜④の具体例としてそれぞれどんな情報が想定されるか，日本語で3つ書き出しなさい。

（例）ご注文いただきました修理が完了したことを報告いたします。
　　　具体例：1) 注文のあった修理の内容
　　　　　　　2) 修理完了日
　　　　　　　3) 修理の請求額および請求方法

① このメールをもちまして，Kosen港を出港したことを報告いたします。

　　1) _____
　　2) _____
　　3) _____

② このメールをもちまして，到着が1日遅れることを通知いたします。

　　1) _____
　　2) _____
　　3) _____

練習H　以下のメール／手紙のまとめ文をp.175を参考にして英語に訳しなさい。

① この手紙に同封しました請求書（invoice）をどうぞご査収ください。

　訳 Please find _____

② 何かこの件に関してご質問やご不明な点がありましたら，いつでもご連絡ください。

🗒訳 _____

③ 修理がいつ完了するかお知らせいただけると幸いです。

🗒訳 _____

④ 添付資料をご覧ください。

🗒訳 _____

◇リーディング・海技試験英語対策

次の英文を日本文になおしなさい。

We wish you to report that on our arrival here at 8:30 p.m. 20th May, and after completion of discharge of whole cargo, we underwent a survey of Lloyd's surveryor, Mr. Eliot, and in accordance with his recommendation we ordered necessary repairs by Messrs. B. & V Dock Yards.

After repairs were finished on 23rd inst., to the satisfaction of Mr. Eliot, he granted the seaworthy certificate for our homeward voyage.

We have the pleasure to enclose our approximate damage list, sketch of damaged part, and copies of each survey report.

The bill of repairs has not yet reached us, but we will send it to you as soon as we receive it.

(『海技士2N徹底攻略問題集』より)

《語　注》

1. 23rd inst.：inst.はinstantの略で，「今月の」の意味
2. to the satisfaction of：〜の満足するようなやり方で

Story 19　How Do You Read Me?

楽しかったハワイでの休暇もあっという間に過ぎ，また忙しいけれど充実した日常に戻ってきたMioさん。今朝はまた通常どおり，6時半に起床，8時15分に出社です。前直より引き継ぎを受け，最初の業務はKosen港に入ってくる外航船とのVHF通信です。なんとこの船は，同じ高等専門学校を卒業したKouくんとKaiくんが乗船している大型自動車船です。

◎海しごと＆海事知識

入港予定の港に近づいた船舶はその港のポートラジオと連絡をとり，安全な入港に向けた体制をとる。細かい規定は港ごとに異なるが，ここではその一例をみてみる。

① 入港

通報する時間帯	船舶から通報する情報	ポートラジオから通報する情報
入港2時間前	P/S到着予定時刻	パイロット着次第乗船または必要な時間調整
入港1時間前において変更などが生じた場合	P/S到着予定時刻の変更	パイロット着次第乗船または必要な時間調整
B/W約20分前	B/W通過予定時刻	港内他船行き会い情報

P/S：pilot station（パイロットステーション，水先人との合流水域）
B/W：breakwater（防波堤）

② 出港

通報する時間帯	船舶から通報する情報	ポートラジオから通報する情報
離岸前	離岸予定時刻	港内他船行き会い情報
離岸予定時刻に変更が生じた場合	離岸予定時刻	港内他船行き会い情報

また，上記以外にも，港や船舶の保安レベルや係留場所などの情報も交わされる。

◉ダイアログ

Ship : Kosen port radio, this is the Maritime Dream.
Mio : The Maritime Dream, this is Kosen port radio. Change to channel 11.
Ship : Kosen port radio, this is the Maritime Dream.
Mio : The Maritime Dream, this is Kosen port radio. How do you read me?
Ship : I read you excellent.
Mio : The Maritime Dream, message go ahead.
Ship : I am passing Hope Suido traffic route. My ETA at Kosen Wan Light buoy No. 4 is 1030 hours local time, over.
Mio : Roger, your ETA at Kosen Wan Light buoy No.4 is 1030 hours. No inbound and outbound vessels at the time of your ETA. You may proceed to your berth directly. Your berth is Kibo Pier. Berth starboard side alongside, over.
Ship : Roger.
Mio : Please call me again after you get alongside.

語彙

1. pass : 〔動詞〕 〜を通過する
2. traffic route : 航路
3. ETA (Estimated Time of Arrival) : 到着予定時刻

4. light buoy：灯浮標，ライトブイ
5. local time：現地時間
6. bound：〔形容詞〕～行きの
7. inbound：〔形容詞〕入ってくる，本国行きの ⇒ inbound vessel：入港船
8. outbound：〔形容詞〕出て行く，外国行きの ⇒ outbound vessel：出港船
9. proceed：〔動詞〕進む，前進する
10. berth：〔動詞〕停泊する／〔名詞〕係留場所，停泊位置
11. starboard：〔名詞〕右舷／〔形容詞〕右舷の
12. alongside：〔副詞〕横付けに，舷側に ⇒ get alongside：着岸する

フォーカスする文法／表現

【英語ワザ-1】海仕事関連語句や表現：無線英語

◆無線の感度に関する表現
- How do you read me?　感度はどうですか？

  ＜応答例＞

  感度は最悪です：I read you bad

  感度は悪いです：I read you poor

  感度は普通です：I read you fair

  感度は良好です：I read you good

  感度は最高です：I read you excellent

- This is...　こちら○○です。

  This is the Maritime Dream, over.

◆無線会話の区切りの表現

over：VHS無線電話は一方通行の会話になるので，会話終了のサインとして"over"や「どうぞ」をいう

roger ／ copy：了解した（無線で使われる英語）

◆その他の頻出単語／表現

berth

inbound ／ outbound

starboard：右舷 / port：左舷（荷の上げ下ろしを左舷でする）
alongside：接舷して，〜の舷側に
ETA：到着予定時刻
local time

練習A　ダイアログからの抜粋です。①〜④の対話文を2回音読しなさい。音読が終わったら，終了マーク○を赤く塗りなさい。

　　　　　　　　　　　　　　　　○音読1回目終了　　○音読2回目終了

① Mio：How do you read me?
　 Ship：I read you excellent/good.
② Mio：The Maritime Dream, message go ahead.
　 Ship：I am passing Hope Suido traffic route.
③ Ship：My ETA at Kosen Wan Light buoy No.4 is 1030 hours local time, over.
　 Mio：Roger.
④ Mio：No inbound and outbound vessels at the time of your ETA.
　　　　You may proceed to your berth directly. Your berth is Kibo Pier.
　　　　Berth starboard side alongside, over.
　 Ship：Roger.

練習B　日本文を参考に空欄に適切な単語を入れなさい。
① 本船は左舷係留の予定です。

　　We will berth (　　　　　　　) side (　　　　　　　　).

② 右舷錨を投錨する。

　　We will let go (　　　　　　　) anchor.

③ 10時の方向に出港船です。

　　There is an (　　　　　　　) vessel in 10 o'clock direction.

④ 本船の到着予定時刻は現地時間午前8時です。

　　Our (　　　　　　) is 0800 hours (　　　　　　　)
　　(　　　　　　).

【英語ワザ-2】 復習：海事通信における数字の読み方
◆ SMCP（標準海事通信用語集）に基づいて，海事通信では数字は，操舵号令を除き一桁ずつ読む
　（例1）150：one five zero
　（例2）2.5：two point five ／ two decimal five

練習C　　次の数字の読み方を英語で書き，1回音読しなさい。
（例）Change to channel 11.【one one】
① Stand by on VHF 16.　　　　　　　【　　　　　　　　　　】

② Charted course is 055 course.　　　【　　　　　　　　　　】

③ Magnetic compass course is 057 course.【　　　　　　　　　　】

④ Dry temperature is 25.0 ℃.　　　　【　　　　　　　　　　】

⑤ Visivility is 15 miles.　　　　　　【　　　　　　　　　　】

【英語ワザ-3】 通常の数字の読み方（計算と大きな桁の数字の読み方）
◆ 英語で簡単な計算を言えるように
・以下の動詞は受動態で用いられることが多い。

add：足す（⇒ addition：〔名詞〕足し算）
　（例）Ten <u>added to</u> five is fifteen.　（5に10を足すと15）
subtract：引く（⇒ subtraction：〔名詞〕引き算）
　（例）Fifty five <u>subtracted from</u> one hundred is forty five.
　　　（100から55を引くと45）
multiply：掛ける（⇒ multiplication：〔名詞〕掛け算）
　（例）Ten <u>multiplied by</u> five is fifty.　（10に5を掛けると50）
divide：割る（⇒ division：〔名詞〕割り算）
　（例）Five hundred <u>divided by</u> one hundred is five.　（500を100で割ると5）

- 他の表現

  times：〜倍する　　　　　　　（例）Three times five is fifteen.
  equal：等しい，イコール〜だ　（例）Five times two equals ten.
  plus（前置詞的用法）　　　　　（例）Three plus two equals five.
  minus（前置詞的用法）　　　　（例）Four minus three is one.

練習D　次の数式の読み方を【　】の単語を使って英語で書きなさい。その後，その式と答えを1回音読しなさい。

① 15 + 35 = 50　【plus】

② 8 + 12 = 20　【add】

③ 83 − 27 = 56　【minus】

④ 83 − 27 = 56　【subtract】

⑤ 32 × 6 = 192　【times】

⑥ 32 × 6 = 192　【multiply】

⑦ 72 ÷ 8 = 9　【divide】_____

◆3桁までの数字の読み方
- 基本形
  （例）123：one hundred and twenty three
- 口語形
  口語ではandが省略されることが多い。
  （例）123：one hudred twenty three

◆4桁以上の数字の読み方
大きい桁の数字はまず3桁ごとにカンマを振ってみる。
このカンマの位置で使われる語がポイントとなる（以下，カンマの位置を／で示す）。
百の位のすぐ左のカンマの位置ではthousandが使われる。その右にある3桁はまず無視し，千の位を読む。その後に3桁の数字を読む。
　（例）1,234：one thousand / two hundred (and) thirty four
もう1桁増えた場合，カンマの左側の数字を読み，そこにthousandを足す。その後にカンマの右側の3桁の数字を読む。もう1桁増えても同じ。
　（例1）12,345　：twelve thousand / three hundred forty five
　（例2）123,456：one hundred twenty three thousand / four hundred fifty six
2つ目のカンマの位置ではmillionが使われる。読み方はこれまでと同じ。まず右側の6桁は無視する。
　（例1）1,234,567：　one million / two hundred thirty four thousand /
　　　　　　　　　　　five hundred sixty seven
　（例2）12,345,678：twelve million / three hundred fourty five thousand /
　　　　　　　　　　　six hundred seventy eight
3つ目のカンマの位置ではbillion，4つ目のカンマの位置ではtrillionが使われる。読み方は上記と同じルール。

英語数字の数え方

数字	英語表記	日本語表記
1	one	一
10	ten	十
100	one hundred	百
1,000	one thousand	一千
10,000	ten thousand	一万
100,000	one hundred thousand	十万
1,000,000	one million	百万
10,000,000	ten million	一千万
100,000,000	one hundred million	一億
1,000,000,000	one billion	十億

◇リーディング・海技試験英語対策

以下の英文を日本語になおしなさい。

The IMO SMCP includes phrases which have been developed to cover the most important safety-related fields of verbal shore-to-ship (and vice versa), ship-to-ship and on-board communications. The aim is to get round the problem of language barriers at sea and avoid misunderstandings which can cause accidents.

The IMO SMCP builds on a basic knowledge of English and has been drafted in a simplified version of maritime English. It includes phrases for use in routine situations such as berthing / as well as standard phrases and responses for use in emergency situations.

（IMOホームページhttp://www.imo.org/OurWork/Safety/Navigation/Pages/
StandardMarineCommunicationPhrases.aspxより）

《語　注》
1. verbal：〔形容詞〕口頭の
2. shore-to-ship：陸から船への
3. vice versa：逆もまた同様に
4. ship-to-ship：船舶間の
5. get round：〜を避ける，避けて通る
6. avoid：〔動詞〕避ける，回避する
7. draft：〔動詞〕作成する
8. simplified：〔形容詞〕簡易化した，簡単にした

Story 20 Let's Toast!

同じ高等専門学校を卒業したKouくん，Kaiくん，そしてMioさん。長い航海を終え下船したKouくんとKaiくん，そしてKosen港で仕事をするMioさんは久しぶりの再会を居酒屋で祝うことにしました。お世話になった三等航海士そして三等機関士の先輩方も一緒です。

◎海しごと＆海事知識

＜外航船舶職員の休日・休暇＞
乗船勤務している船員の休日・休暇は連続してまとめて取得されている。一般会社員の連続休暇がゴールデンウイークの10日程度であるのに比べ，6カ月乗船後の2カ月程度の休暇取得など，長い連続休暇となっている。乗船した場合，次の休暇まで帰宅することは困難であるが，この長期休暇を利用して旅行をはじめさまざまな活動ができる。このように「長期休暇が取れること」は船員職業の大きな魅力のひとつである。以下，船員として勤務した場合の休日・休暇について解説する。

年間の休日と休暇の付与およびその取り扱いなどは労働協約で定められている。労働協約上，陸上休暇は年間120日（有給休暇25労働日を含む）とされており，この他に乗船中は週1日の船内休日がある。また，本人の結婚，近親者の死亡に対応した特別休暇の付与など，船員の特殊事情を配慮した休暇制度が定められている。乗船期間は，船員法では12カ月を上回らないこととなっている。乗船した船舶への習熟に要する期間の必要性から3カ月以上の乗船が求められること，休暇取得の要求と社会的な奨励などから，6カ月乗船後の2カ月程度の休暇取得の乗船・休暇サイクルが望ましいとされている。現在は日本人職員の不足などにより，陸上休暇を消化できずに次の乗船を指示されることもある。その場合は一定の基準で「休暇買い上げ」の規程があり，休暇の代わりに金銭的補償が与えられている。

（『船しごと、海しごと。』より）

◉ダイアログ

Kou : Let's toast to our wonderful career in the maritime industry!

All : Cheers!

Kai : Thank you so much for everything. I learned so much from you. Everyone, he is the one who taught me everything on my first assignment.

3/E : It was my pleasure working with you. I think you are going to be a great engineer.

Kai : You are so kind, but thank you.

3/O : So, how was your first voyage, Kou?

Kou : At the beginning, I was not sure if I could survive such a long voyage. Especially, I was worried about my English.

3/O : I think you did a good job for the first time.

Mio : And you, Kai? How was your first voyage?

Kai : I too was worried about my English. But the crew really tried to understand my broken English. And they were really kind to me. I got to know some of them very well.

3/E : I think that was your key to success. You tried communicating with them.

Kou : Hey Mio, I was so surprised to hear your voice over the VHF!

Mio : I know me too! At first, I wasn't sure if it was you. But the voice sounded so familiar. You sounded too good for somebody who flunked every other English test in college!

All : (Laughter)

語　彙

1. toast to：〔動詞〕〜に祝杯をあげる
2. maritime industry：海運業（界）
3. survive：〔動詞〕なんとかやっていける，生き残る
4. flunk：〔動詞〕落第する

フォーカスする文法／表現

【英語ワザ-1】ifの用法「〜かどうか」
◆Ifは「もし〜だったら」の仮定／条件を表すだけでなく，「〜かどうか」の意の間接疑問文をつくるのに用いられる
◆疑問文には直接疑問文と間接疑問文がある
- 直接疑問文

直接疑問文は「質問」を「直接」問いかける文である。
直接疑問文には疑問詞の付くものと，付かないものがある。
（例1）<u>What</u> time is it?（疑問詞あり）「いま何時ですか？」
（例2）Do you like this?（疑問詞なし）「これ好きですか？」
- 間接疑問文

間接疑問文は「質問」に「間接」的な語句が付けられている。
間接疑問文では，疑問文内の語順が肯定文へと変化する。
　　Do you know ＋ "What time <u>is it</u>?" ＝ Do you know what time <u>it is</u>?
　　　　　　　　　　　　　　　　　　　　　いま何時かわかりますか？
付け加えられる語句の時制に合わせて，疑問文の時制を変える。
　　I was not sure ＋ "<u>Is</u> it him?" ＝ I was not sure if it <u>was</u> him.
間接疑問文にした場合に，疑問文の代名詞が変わることがある。
　　I was not sure ＋ "Is it <u>him</u> (Kou)?" ＝ I was not sure if it was <u>you</u>.
　　　　　　　　　　　　　　　　　　（Kouに話しかけているので）
疑問詞のない疑問文には if または whether を使い「〜かどうか」という意味にする。

I want to know +"Do you like this?"= I want to know <u>if</u> you like this.
　　　　　　　　　　　これ好きかどうか知りたいのです。

◆心のなかのつぶやきも疑問文
直接誰かに投げかけるだけでなく，心のなかでのつぶやきも疑問文。
不安に思いながら Can I survive such a long voyage?「長い航海やっていけるかな？」と心でつぶやいたこと。そのような過去の心のつぶやきを振り返って I was not sure（当時は自信がなかった）if I could survive such a long voyage. （こんな長い航海やっていけるかどうか）と間接疑問文にすることができる。
また，VHFの相手の声を聞きながら Is it him?「（ひょっとして）彼かな？」と心のなかで思っていたことを振り返って，後日その当人に，I was not sure（そのときは確信が持てなかった）if it was you.（あなたかどうか）とすることもできる。

練習A　以下の直接疑問文を，【　】のなかの語句を用いて間接疑問文にしなさい。
① 【do you know】"Who is on watch now?"

答_____

② 【Please explain】"Why are they late?"

答_____

③ 【Could you please show me】"How do you clean the strainer?"

答_____

④ 【I am not sure】"Is this yours?"

答_____

⑤【The captain asked】"Do you understand?"

答 _____

⑥【They don't know】"What is yose-nabe?"

答 _____

⑦【I ask the ship】"What is your ETA?"

答 _____

【英語ワザ-2】 お礼に対する返事

◆ダイレクトに伝える基本形

A：You're welcome.
B：No problem.（たいしたことないよ，というニュアンス）
　　Anytime!　（またいつでも言ってね，というニュアンス）
　　My pleasure.（何かをしてあげたが，相手より自分のほうが喜んでいるというニュアンス）

◆フォーマルな表現

pleasure〔名詞〕：喜び

（例）A：Thank you so much for teaching me many things.
　　　　たくさんのことを教えてくださってありがとうございました。
　　　B：It was my pleasure working with you.
　　　　一緒に仕事ができてとてもうれしかったよ。

honor〔名詞〕：光栄

（例）A：I really appreciate all that you have taught me.
　　　　教えてくださったすべてのことに心から感謝しています。
　　　B：It was my honor to have you as my student.
　　　　あなたを教え子として持つことができてとても光栄だったよ。

練習B　以下の状況でのお礼の言葉に最も適切な返事を㋐, ㋑, ㋒から選びなさい。

① 状況：ドアを開けてあげる
　A：Thank you.

　B：_____

② 状況：卒業式で先生にお礼を伝える
　A：Thank you so much for everything. You taught us so much.

　B：_____

③ 状況：フィリピン人クルーに日本語を教えてあげる
　A：Thank you for teaching me Japanese.

　B：_____

㋐ Anytime!
㋑ It was my honor to have you as my student.
㋒ No problem.

【英語ワザ-3】 感謝の気持ちを伝える表現
◆ダイレクトに伝える基本形：Thank you
　Thank you.
　Thank you for 名詞／動詞ing.
　（例）Thank you for teaching me many things.
◆感謝の表現に加え，なぜ感謝しているのか，どれだけ感謝しているのかを一言加える
　（例1）Thank you so much. I learned so much from you.
　（例2）Thank you for everything. I could not have done this without you.
　（例3）Thank you for coming tonight. I really wanted you to meet my family.

◆フォーマルな表現：appreciate

appreciate：〔動詞〕～をありがたく思う，感謝する

（例1）I appreciate your kindness.
　　　 ご親切に心から感謝いたします。

（例2）I deeply appreciate what you have done for me.
　　　 ご厚情に感謝いたします。

appreciation：〔名詞〕感謝

- スピーチ／挨拶などによく使われる
  （例）I would like to express my appreciation for what you have done for me.
  　　　本当によくしていただき，心より感謝の意を表したいと思います。

- 贈りものを渡すとき
  （例）Please take this as a symbol of my appreciation.
  　　　どうぞ感謝の気持としてこれを受け取ってください。

appreciateはお願いをする表現としても使われ，「～していただけたら（条件）感謝します」＝「～していただけたら幸いです」という意味となる。

（例）I would appreciate it if you could give us a permission.
　　　許可をいただけると幸いです。

（例）Your cooperation would be greatly appreciated.
　　　ご協力いただければ非常に幸いに思います。

練習C　　下記の英文を2回音読しなさい。音読が終わったら下記の終了マーク〇を赤く塗りなさい。

〇音読1回目終了　　〇音読2回目終了

① Thank you so much for everything.
　 いろいろとありがとうございました。

② Thank you very much for visiting our booth.
　 私どもの展示ブースに来ていただき，ありがとうございます。

③ Thank you very much for what you have done for me.
　 ご厚情に感謝いたします。

④ I greatly appreciate your kindness.

ご親切に心から感謝いたします。
⑤ I appreciate what you have done for me.
「お世話になりました。」のていねいな表現
⑥ Thank you for your kindness.
「お世話になりました。」の略式表現
⑦ I would appreciate it if you could permit us to get off the ship immediately.
ただちに下船の許可をいただければ幸いです。

練習D　感謝の気持ちを伝えたい人を2人思い浮かべ，英語でお礼を述べなさい。その際，なぜ感謝しているのか，何をしてくれたことに感謝しているのかを表現すること。

① 誰に感謝をしているのか：＿＿＿＿＿＿＿＿＿＿＿＿＿＿＿＿＿＿＿＿

　感謝の表現：＿＿＿＿＿＿＿＿＿＿＿＿＿＿＿＿＿＿＿＿＿＿＿＿＿＿

② 誰に感謝をしているのか：＿＿＿＿＿＿＿＿＿＿＿＿＿＿＿＿＿＿＿＿

　感謝の表現：＿＿＿＿＿＿＿＿＿＿＿＿＿＿＿＿＿＿＿＿＿＿＿＿＿＿

練習E　下船することが決まりました。お世話になった方にお礼の気持ちを口頭で伝えたいと思います。練習CやDの英文を利用して，5～6文程度で書いてください。

答

◇リーディング・海技試験英語対策

以下の英文を日本語になおしなさい。

Guideline B2.4.2 — Taking of annual leave

1. The time at which annual leave is to be taken should, unless it is fixed by regulation, collective agreement, arbitration award or other means consistent with national practice, be determined by the shipowner after consultation and, as far as possible, in agreement with the seafarers concerned or their representatives.
2. Seafarers should in principle have the right to take annual leave in the place with which they have a substantial connection, which would normally be the same as the place to which they are entitled to be repatriated. Seafarers should not be required without their consent to take annual leave due to them in another place except under the provisions of a seafarers' employment agreement or of national laws or regulations.
3. If seafarers are required to take their annual leave from a place other than that permitted by paragraph 2 of this Guideline, they should be entitled to free transportation to the place where they were engaged or recruited, whichever is nearer their home; subsistence and other costs directly involved should be for the account of the shipowner; the travel time involved should not be deducted from the annual leave with pay due to the seafarer.
4. A seafarer taking annual leave should be recalled only in cases of extreme emergency and with the seafarer's consent.

（MARITIME LABOUR CONVENTION, 2006（海事労働条約）http://www.ilo.org/wcmsp5/groups/public/@ed_norm/@normes/documents/normativeinstrument/wcms_090250.pdfより）

《語　注》
1. annual leave：年次休暇
2. fix：〔動詞〕固定する，決定する

Story 20 Let's Toast!

3. regulation：〔名詞〕条例，規制
4. collective agreement：団体協約
5. arbitration award：仲裁裁定
6. means：〔名詞〕方法
7. consistent with：矛盾のない，一貫した
8. national practice：国の慣例／慣行
9. consultation：〔名詞〕協議，相談
10. as far as possible：できる限り
11. seafarers concerned：関係する船員
12. representative：〔名詞〕代理人
13. in principle：原則として
14. substantial：〔形容詞〕実質的な，かなりの
15. be entitled to：〜する資格がある
16. be repatriated：帰還する，本国に送還される
17. consent：〔名詞〕承諾，同意
18. provisions：〔名詞〕規定，条項
19. employment agreement：雇用契約
20. permit：〔動詞〕許可する，許す，認める
21. engaged：雇われる
22. recruited：採用される
23. subsistence cost：（必要最低限の）生活費
24. for the account of：〜の負担で
25. deduct：〔動詞〕差し引く，控除する，天引きする
26. leave with pay：有給休暇
27. due to：〜に当然支払われるべき
28. recall：〔動詞〕取り消す，呼び戻す
29. extreme emergency：極度の緊急事態

練習の解答例／リーディングの訳

Story 1

練習A
① I am going to talk about the types of ships we operate.
Types of ships we operate
② 4つ
③ we operate various kinds of ships

練習B
① Toolbox meeting / A toolbox meeting is a short safety talk.
② 3つ
③ a specific safety message

練習C
① これから私たちが乗船する船
② 私が午前中に行うべき最初の作業
③ あなたが入力するデータ
④ 鉄心の可動距離
⑤ それぞれお好みの火の通った具材
⑥ あなたがつねに持ち歩かなくてはならない数多くの重要書類

練習D
① Container ships are designed (to carry goods in truck-size containers).
② A valve is designed (to control currents of liquids and gases).
③ Bulk carriers are built (to carry non-packed materials).
④ The Yellow Card (is known as the Yellow Fever Card).
⑤ Night work of seafarers under the age of 18 (shall be prohibited).
⑥ The procedure (must not be used) at the same time by multiple aircrafts or by multiple vessels.
⑦ We must make sure that (goods are delivered in) a timely manner.
⑧ Radio transmission (should be made as soon) as possible.

リーディング

ばら積み船の安全

ばら積み船は穀物や石炭そして鉄鉱石など無包装の物資を大量に運ぶために1950年代に開発された。約5000隻のばら積み船が現在世界中で運航しており，国際物資輸送においてなくてはならないサービスを提供している。

ばら積み船の運航者はこの種類の船の運航に特有の安全上の注意を理解していな

ければならない。航海中に貨物がずれて船の安定性に支障を来たさないように，貨物の積み込みは細心の注意をもって行われなければならない。大型のハッチカバーは水を漏らさず，しっかり固定されていなければならない。

Story 2
練習A
① タンカーは(石油)や(化学薬品)のような液体を輸送するために設計されている。
② ばら積み船は(石炭)や(砂利)のような梱包されていない原料を運ぶためにつくられている。
③ 私たちは，(白菜)，(きのこ)，(ネギ)のような多くの野菜，豆腐や肉を入れます。
④ 船員手帳は，(保険)，(有給休暇)，(健康診断書)，そして船員としての雇入履歴など，船員として重要な情報を(含みます)。
⑤ 船員手帳は，(名前)，(生年月日)，そして(住所)のような，あなたの基本的な情報を(含みます)。
⑥ 私は(天候)あるいは港での(運航条件)などのような重要な情報をその船に(伝えます)。
⑦ 三等航海士は(消火器)や(救命艇)など非常時の備品について(担当をしています)。
⑧ たとえば(発煙浮信号)や(無線標識)のような適切なマーカーがその場所に落とされることになっている。
⑨ 海上での生存者は，可能であれば，たとえば2本の(巻き綱)を使って，片方は脇の下，もう一方は膝の下に入れて，水平の位置で(引き上げられる)べきである。

練習B
① I am not good at English, and so I <u>am worried about keeping</u> the logbook in English.
② I <u>feel nervous about communicating with</u> the crew in English, but I am ready to go.
③ This is my first watchstanding, so I <u>got excited at my first duty</u> and woke up early in the morning.
④ I <u>got very surprised to hear</u> your voice over the VHF!

練習C
① あなたは三等機関士として(務めます)。ストレーナを(検査)して(必要)ならばそれを(清掃)しなさい。
② 航海日誌は(衝突)や事故などを(調査)するときに使われる重要な(記録)とし

て（役に立つ）。
③ 私は，よせ鍋を（振る舞う）計画をしています。

リーディング
船員の訓練及び資格証明並びに当直の基準に関する国際条約

1978年のSTCW条約は国際的なレベルにある船員の訓練および資格証明ならびに当直の基本的な資格を最初に定めた条約である。それ以前は，士官および部員の訓練，資格証明，当直の基準は各国政府が制定しており，その制定の際には通常，他の国の慣行に照らし合わせることはなかった。その結果，海運業はもっとも国際的な産業であるにも関わらず，基準や手順が国によって大幅に異なるという状況に至った。この条約は船員の訓練および資格証明ならびに当直の最低基準を定めていて，加盟国はそれを満たすかまたはその基準以上でなくてはならない。

Story 3

練習A
① あなたは（次席三等航海士）として我々の（自動車専用船）のひとつに（任命される）でしょう。
② （できる限り）学びなさい。というのは2度目の（任務）で，あなたは三等航海士として自動車船に乗船するからです。
③ およそ（1週間で），あなたは最初の仕事の（任務）で私たちの船に（乗船する）でしょう。
④ 三等航海士は（消火設備），（救命艇），そしてその他のさまざまな（緊急時）の備品の（担当をする）。
⑤ あなたはこれらの（装備）が（良好な状態）であるということを確認する（責任がある）。
⑥ あなたは，三等機関士のもうひとつの重要な（責任）を（忘れてはならない）。

練習B
① This is your first (assignment) on the ship.
② As a third officer on this ship, I (am) (responsible) for your training.
③ This book talks about various roles and (responsibilities) of deck officers and engineers on our ships.
④ The third officer will (train) you on the daily (responsibilities) and tasks onboard the ship.
⑤ It is the (responsibility) of the shipping agent to (take) a new officer or engineer to the ship.

リーディング
訓練

研修中の航海士は海上でずっと続けられる実地研修と体験を包括的に記録するための訓練記録簿を与えられなければならない。この訓練記録簿は行われるべき作業や任務とその終了までの進展についての詳細な情報を提供するように明確に記述されるべきである。正式に訓練が終わると，訓練記録簿は海上訓練の構造化されたプログラムを修了したというまたとない証拠となり，それは証明書の発行に向けた能力の評価において考慮される。

Story 4

練習A

Are you Mr. E?

You must be Mr. B.

練習B

① Nice to meet you.

② By the way, how was your first assignment, Mr. F?

練習C

① I was worried about my English.

② I am happy to hear that.

③ I was also worried about my English.

④ I was often helped by the crew on the ship, and I got comfortable.

練習D

① <u>Hi, are you Mr. B</u>?

② <u>Yes, you must be A</u>.

③ <u>Yes, I am A</u>. <u>It is good to meet you, Mr. B</u>.

④ <u>Good to meet you too, Mr. A</u>. <u>By the way, how was your voyage</u>?

⑤ It was my <u>first assignment</u>. At the beginning, I <u>got nervous</u>. But <u>the crew</u> was very kind to me. I <u>was helped</u> by them. <u>How was yours</u>?

⑥ It was the same with me. I was <u>worried about my poor English</u>. But I was <u>ready</u>. <u>The crew members</u> were also ready to talk to me. They are <u>my friends now</u>.

⑦ I tried <u>to communicate</u> with them. I showed them my pictures and I <u>introduced them some Japanese food such as takoyaki and sukiyaki</u>. The pictures on the smart phone <u>often helped</u> my poor English.

練習E
① 私たちは(入港)船の(投錨地)を(確保)する。そして必要ならば，パイロットとタグボートを(手配する)。
② 最後に，すべての(火災検知器のセンサ)にスイッチが入っていることを(確認)しなさい。
③ あなたは，これらの(装置)がよい状態にあることを(確かめる)(責任を負う)。
④ 私たちはまた，品物が予定どおりに(引き渡される)ということを(確認する)ために(船主)と荷送人と連絡を取る。
⑤ あなたが船長の(当直命令簿)を理解しているということを(確認しなさい)。
⑥ 私がするべき最初のことは，港湾区域にはあやしい人や(不審船)がいないということを(確認する)ことです。
⑦ すべての乗組員が確実に港湾境界内の(指定)通路を守るように(しなさい)。
⑧ 船長は，船の(進路)，(速度)と(位置)がつねに厳重に監視されている状態を(確認)しなくてはならない。
⑨ パッキン押えをしたパッキン箱がきれいで，すべての古い(パッキン)が(取り外されている)ということを(確認しなさい)。

リーディング
船舶代理店

船は港にいるときがもっとも忙しい。荷物の揚げ下ろしだけでなく，在庫を確認したり，乗組員の交代があったり，さまざまな検査や修理が行われる。やらなくてはならないことのリストは長い。代理店は水先案内の手配を行い，タグボートや運航整備のスケジュールを立て，税関の手続きの代行や船のための各種注文手配を行う。つまり，停泊中の船舶が必要とすることが代理店を通して調整される。船は24時間昼夜を問わずに出入港し，代理店は船が予定どおりに運航できるよう，関係者すべてと連携作業を行う。代理店業は海運業界を構成するあらゆる仕事に触れることができる刺激的かつ活動的な仕事である。

Story 5
練習A
① いつも<u>正午</u>にお昼休みをとります。
② <u>3時</u>にお茶の<u>休憩</u>をとります。
③ <u>8月5日の午前10時に</u>，ここスペインに到着しました。
④ 会社は<u>8月8日</u>に私たちの船の必要な修理を<u>完了させます</u>。
⑤ 私たちの航海は<u>2013年</u>にKosen港から<u>始まりました</u>。

⑥ そちらのほうでは，冬に雪は降りますか？
⑦ 通常，同じ船の上で半年間訓練を行います。
⑧ 私は夏に3週間，インターンシッププログラムに参加しました。
⑨ 午後3時半から5時まで，すべての機関士は夜のMゼロ運転すなわち機関室無人運転のための準備を行います。
⑩ 昼間勤務のときは通常，9時までに出勤します。
⑪ スペインにおいて必要な修理の発注がされ，8月8日までにはその作業が完了した。
⑫ 予期せぬ問題が起こり，我々の船は8月10日までそこに停泊していなくてはならなかった。

練習B
① 通常，我々は朝7時に起床し，それから朝食を食べる。その後に朝の体操を行う。
② まず最初に濾過器バイパス弁を開けなさい。その後，海水の注水口と放水口弁を閉じなさい。そうしたら，容器の内部の圧力を下げ，カバーをはずしなさい。最後に，再びフタをし，ボルトを交互に閉じなさい。
③ まず，船の配置図を見なさい。この配置図には総員退船部署の位置が示されている。次にこのポスターを見なさい。このポスターは点呼名簿と呼ばれ，船のいたるところに張り出されている。

練習C
① ～することができない
② 容認できない，歓迎しがたい
③ 不平等な，不均等な
④ 珍しい，めったにない，まれな
⑤ 特定の任務を与えられていない

リーディング
ツールボックストーク：
ツールボックストークとは，特定の安全に関する問題に焦点を当てた形式張らないグループ内での話し合いである。これらさまざまなツールは部署内のセーフティ・カルチャー（安全文化）を促進するために日々用いることができる。ツールボックストークには仕事の現場において健康や安全に関しての話し合いを促進するという目的もある。

Story 6
練習A
① これから私たちの日課と私たちが船の上ですべきことについてお話します。
② この仕事をあなたはこれから三等航海士としてやっていくので，私がいうことをよく覚えておくこと。
③ 私が書いたことをちょっと見ていただけないでしょうか？
④ 私たちがすべきことの例を挙げます。

練習B
① As the junior third engineer, what exactly do I do?
② What do you think is the most important thing to work on a ship?

練習C
① 新しい士官を（空港で）出迎え，彼を船に連れてくることは私たちの会社にとって重要な仕事です。
② クルーと良い関係を築くことはもっとも重要なことです。
③ すべてのクルーが各緊急装置の使い方を熟知していると確認することがあなたの責任です。
④ 当直を引き渡すことと引き継ぐことはとても重要です。
⑤ 間違いの上に二重線を引いて訂正しなさい。
⑥ すべての積み荷を陸揚げしたあと，私たちはマリーン社のスミス氏による検査を受けた。

リーディング
互いに話をし，ともに笑い，そして，重要なのが，ともにジョークを言い合えるクルーは，さまざまな国籍が混ざり合っていようが，安全かつ楽しく仕事をすることができる。共通言語でコミュニケーションをとる能力は，国籍そして人数に関係なく，多国籍なクルーが成功を決める重要な要素である。船員がよりお互いを理解できればできるほど，たんに効率の良い安全な船の運航だけでなく，良い人間関係と仕事上の関係が構築できる楽しい船の運航が可能となる。

Story 7
練習A
① close ② relieve ③ dismantle ④ put ⑤ tighten ⑥ check
練習B
① 貨物を受け取る船と（連絡をとり），そして提案された作業の（詳細な計画を立

② （係船索），（投索／ヒービングライン），（補助索／メッセンジャーロープ），（ストッパー），（フェンダー）などを（配置しなさい）。
③ 錨が使えそうな水域にあるならば，（その使用に備えてスタンバイさせなさい）。
④ 安全面にとくに触れながら，作業に関して（航海士そしてクルーたちに簡潔な指示を与えなさい）。
⑤ もし可能ならば，（適切な安全チェックリストの項目をすべて確認しなさい）。

練習C
① 持ち物リストを一緒に見ていきましょう。
② イエローカードを所持することを忘れないように。
④ 当直任務がどのように交代されるのかよく見てメモをとっておくように。
⑥ 人がなかにいる場合は，必ずもう1人のひとが必ずボイラーのそばで待機しているようにしなさい。

練習D
①：ウ　②：オ　③：エ　④：ア　⑤：キ　⑥：カ　⑦：イ

リーディング
海水ポンプシステムで使われているボルトやネジは錆びやすい傾向にあり，取り外す際にひねることで壊れてしまうことがある。とくに壊れた植え込みボルトが土台に残ってしまった場合は厄介だ。ポンプを底板に固定している植え込みボルトが壊れ，ポンプの本体を再び設置しなくてはならなくなり，その作業に3日を要したという人もいる。狭い空間でのそのような作業は多大な時間を必要とする。

Story 8

練習A
① in　② on　③ in　④ on　⑤ in　⑥ at　⑦ in
⑧ in または at（管制塔のなかでというイメージの場合は in，管制塔の具体的な場所で作業していることをイメージしている場合は at で，両方可能）

練習B
① with　② with　③ without　④ with
⑤ without　⑥ within　⑦ within　⑧ within

練習C
① 私は長期の航海にでるときはいつも家族の写真を持っています。
② 何か異常に気づいたときはいつも船長に報告しなさい。
③ 新しく乗船してきた人がいるときはいつも，船舶の安全と救命装置だけでなく

非常時の対応についての入門指導を準備するのが三等航海士の仕事です。
④ ライナまたはピストンが交換された場合はいつも，ピストンとライナまたはシリンダーのすき間は検査されねばならない。
⑤ 適切と判断される場合はいつも，これらの検査は記録されるべきである。

練習D
① when/as　　② after/before　　③ while　　④ if
⑤ in case　　⑥ although/though　　⑦ because/since/as　　⑧ so that

練習E
① このリストは緊急の種類についての情報と，さまざまな緊急事態の場合に従うべき指示を含む。
② このリストは全乗組員の共通集合場所，救命艇の任務，非常時の場合の各人および部署の任務を含む。
③ 陸揚げの港で，積荷が荷受人にとって満足のいかない状態だった場合，参考のため上記に言及されている実際の状況や周囲環境について書き留めてください。

リーディング
点呼名簿および非常時の指示
(1) この規則は，すべての船舶に適用する。
(2) 非常の際に従うべき明確な指示はすべての乗船者に与えるべきである。
(3) 第37規則の規定に適合する点呼名簿は，船橋，機関室，乗組員の居住区域を含む船内全域の目につきやすい場所に掲示されるべきである。
(4) 次の事項を旅客に知らせるため，適当な言語による説明図および指示は，客室に掲示し，かつ，集合場所および他の旅客区域に目につきやすいように掲示されるべきである。
    (a) 集合場所
    (b) 非常の際にとらなければならない不可欠な行動
    (c) 救命胴衣の着用方法

Story 9

練習A
① 新しいテーブルクロスをこのテーブルに広げてください。
② 間違いの上に二重線を引いて，訂正しなさい。
③ 16歳未満の人の雇用は禁じられている。
④ 18歳未満の船員の夜間作業は禁じられている。
⑤ 今日は荷物につめるもののリストを一緒に確認していきましょう。

⑥ 航海当直の引き継ぎの流れを説明してあげますよ。
⑦ 船橋航海当直の引き渡しと引き継ぎは船を安全に運航するために重要である。
⑧ どのように当直業務が引き渡されるのかをよく観察し，メモしておくように。
⑨ 機関士と部員は安全注意事項を復唱するために集合します。

練習B
① 当直担当士官は次の当直担当士官に安全な航行に関連性のある情報を簡潔に伝えます。
② 大学の英語のテストでしょっちゅう落第点をとった人にしては，よくできてたよ。
③ イエローカードは，あなたが黄熱病に対してワクチンを打ったことを示す書類です。
④ Kosen港を出入港する船と連絡を取ります。
⑤ 訂正を加えられた法的文書は，変更をした人の捺印がなくてはならない。

練習C
① イエローカードを所持することを忘れないように。
② ストレーナーのカバーから漏れがないか確認することを忘れないように。
③ 船の安全と救命設備に関してのオリエンテーションを行うことは三等航海士の仕事です。
④ 必要に応じてそのリストを更新することは安全担当者の責任です。
⑤ しっかりと見張ることが最も大切な仕事です。
⑥ 私が最初にやるべきことは，港に不審者や不審船がいないかを確認することです。

練習D
① 荷物につめるもののリストを一緒に確認していきましょう。
② このリストには非常時の種類に関する情報と従うべき指示が含まれています。
③ （近況について）聞きたいことがたくさんあります。

練習E
① タンカーは石油や化学薬品などの液体を輸送するために設計されている。
② ばら積み船は石炭や砂利のような梱包されていない原料を運ぶためにつくられている。
③ 自動車専用船は車を輸送するためにつくられている。
④ コンテナ船はトラックサイズのコンテナに入った物資を運ぶために設計されている。
⑤ 航海士，機関士そしてクルーが職場でよい関係性を保つために，われわれは努力しなくてはならない。
⑥ 荷物を必ず安全に届けるために，我々は船主と船荷主の両方と連絡をとる。

⑦ 船橋当直は船を安全に運航するためにとても重要なことだ。
⑧ 当直の引き継ぎは,航海に関する情報を確認し,エラーを正すために重要なタイミングだ。
⑨ 我々は衝突を回避するためにこちらに向かっている船舶にいちはやく気づき,適切な行動をとらなくてはならない。
⑩ 娘の誕生日を祝うために家に居られたらいいのにな。

リーディング

航海当直に当たる士官は,搭載している電子航海計器の使用を,その装置の性能や限界を含み(完全に)熟知し,また適切な時にそれぞれの装置を使うべきである。また,音響測深機が貴重な航海計器であることを心に留めておくべきである。

航海当直の責任を負う士官は,視界制限状態にあるかあるいは予想される時はいつでも,また船舶の輻輳水域においてはつねに水域制限に十分配慮しながら,レーダーを使うべきである。

航海当直士官は,レーダーエコーができるだけ早く探知できるように,使用中のレーダーのレンジスケールを,かなり頻繁に変えることを確実に行うべきである。小さいあるいは弱いレーダーエコーは探知されないということも覚えておくべきである。

Story 10

練習A

① Zero five seven degrees
② One one point zero knots / One one decimal zero knots
③ One zero one two point nine hectopascal / One zero one two decimal nine hectopascal
④ Two five point zero degrees Celsius / Two five decimal zero degrees Celsius
⑤ One five miles
⑥ One four zero zero hours

練習B

① 風力は風力階級で2から3に上がりました。
② 現在の位置はコースライン上から2ケーブル左です。
③ 次の変針点まで3マイルです。
④ 風向は北から北東へ変わりました。
⑤ 同航船があります。方位は正船尾,距離は1.5マイルです。
⑥ 1400時に,御前崎灯台を左舷正横後1ポイント,方位340°,距離12マイルに見

ました。

練習C
① We are (on) the charted course (to) Tokyo bay.
② We are sailing 13 miles (off) the coast of the Kii Peninsula.
③ What is the distance (to) the new waypoint?
④ Weather changed (from) b (to) bc.
⑤ 12 miles (to) the next waypoint.
⑥ The vessel 20 degrees on the port bow is (on) the opposite course.
⑦ The vessel on the starboard quarter is (on) the same course.
⑧ Present position is 2 cables right (from) the charted course line.
⑨ Take No. 2 generator (off)-line.
⑩ Put No. 2 generator (on)-line.

練習D
① 左舷／右舷
② 左舷船首／右舷船首
③ ヘッドライン／スターンライン
④ 船首／船尾
⑤ 船尾に／後方に／船首に
⑥ 前方への，前方へ／船尾にある，船尾に，後方に
⑦ 船体中央
⑧ 船腹／真横に
⑨ 前部ブレストライン／後部ブレストライン
⑩ 前部スプリング／後部スプリング
⑪ 船首上甲板
⑫ 左舷船尾側／右舷船尾側
⑬ 船橋，ブリッジ
⑭ センターライン，中心線
⑮ 幅，全幅
⑯ 全長

リーディング
船舶から陸へまたその逆，船舶間，そして船上での航海や安全に関するコミュニケーションは，混乱や間違いを避けるために正確かつ簡潔そしてあいまいさのないものでなければならないため，使用される言語を標準化する必要がある。コミュニケーション上の問題は誤解を生み，船およびその乗組員，ひいては環境に対して

危険をもたらす可能性があるため，多言語を話すクルーを乗せた国際的な貿易船が増加しているという観点からみると，このことは格別重要となってくる。（中略）1978年の船員の訓練及び資格証明並びに当直の基準に関する国際条約（1995年改訂）ではIMO（国際海事機関）のSMCP（標準海事通信用語集）を理解し使える能力が500総トン以上の船の航海当直に責任を持つ士官の資格証明に必要とされている。

Story 11

練習A
① a soup dish　② ingredients　③ time (of arrival and departure)
④ important record　　⑤ past log book entries
⑥ any corrections　　⑦ walkways

練習B
① 深鍋で料理された汁物の料理
② 調理された好きな具材
③ 予定されている到着と出発の時刻
④ 衝突や事故などの調査に使われる重要な記録
⑤ 先輩の士官によって書かれた過去の航海日誌の記録
⑥ 航海日誌に加えられたすべての訂正は
⑦ 港湾内の指定された通路

練習C
① 乗組員は彼らの安全のために用意されているすべての設備の正しい使用法を熟知すべきである。
② サルベージマスターより与えられるあらゆる助言を船長は考慮すべきである。
③ 船の主要な操作に使われるすべての部品は浸水に対して保護されるべきである。
④ カムとはバルブの開閉に使われている回転している円盤の上についている真円ではない突起である。
⑤ 排気弁箱は油圧ジャッキによって固く締められた植え込みボルトとナットのついたシリンダカバーに接続している。
⑥ 3時半から5時まで，すべての機関士は夜間のMゼロ運転すなわち機関室無人運転のための準備を行います。

練習D
① 明日のパーティーではどんな食べ物を出す予定ですか？
② 寄せ鍋を出す予定です。
③ これから，私たちが運航する船の種類についてお話します。

④ まもなく，三等航海士としてあなた自身がすることになります。だから，これから私が話をすることをよく覚えておきなさい。
⑤ 数年たつと，あなたは素晴らしい機関士になると私は思います。

練習E
① じゃあ，今日，新しいのを買うよ。
② 今日，新しいのを買いに行くつもりだよ。
③ すぐに雨が降るよ。
④ 天気予報によれば，明日は雨になるだろう。
⑤ 君はジュニアサードオフィサーとしてKosen港から乗船するだろう。
⑥ これから船に乗ります。
⑦ はい，わかりました。
⑧ 当直を引き渡す前に簡潔に報告します（＝当直交代前の引き継ぎを行います）。

練習F
① We call it (a pot) in English.
② Yosenabe means soup dish (cooked) in a Nabe pot.
③ Yosenabe (is) often (cooked) at the table.
④ We put a lot of vegetables such as (Chinese cabbage), (mushrooms), (green onions), (tofu) and meat of your choice.
⑤ We can pick (any) (cooked) (ingredients) that we like from the pot.
⑥ We can eat them (with) soup or (dip) it in the sauce.

練習G
① What do you call this in Tagalog?
② How do you say "Thank you" in Tagalog?
③ "Unmanned engine room" とはどういう意味ですか？
④ あなたの言葉では "good night" をどういうのですか？

リーディング
安全な航海および機関当直を維持するためには，当直士官は以下を含むブリッジ／エンジンルーム・リソースマネジメントの原則に関する十分な知識を持っていなくてはならない。

1. リソース（資源）の配置，任務，および優先順位決定
2. 効果的なコミュニケーション
3. 明確な意思表示とリーダーシップ
4. 状況認識力の獲得と維持
5. チームの体験への配慮

Story 12

練習A
① I am from Toyama.
② I live in Imizu city.
③ Is it located in the western part of Toyama.
④ It is famous for delicious water and fresh seafood.
⑤ It is located in the northwestern part of Tokyo.

練習B
① Hi, where in the Philippines are you from?
② I am from Mindanao Island.
  It is located in the southern part of Manila.
③ Where in Japan are you from?
④ I am from Toyama.
  It is located to the west of Tokyo, on the main island of Japan.
  Toyama is famous for its nature such as Mt. Tate, water, and seafood.
  I am so glad to meet you.
⑤ Glad to meet you too.
  Does it snow over there?
  I have never seen snow.
⑥ Yes, it snows a lot.
  I will take you to Toyama after this voyage.
⑦ Thank you, you are so kind.

練習E
① 新しいモデルのパソコンを持っていたらいいなあ。
② 英語をうまく話せたらいいのになあ。
③ 日本に地震がなければいいなあ。
④ まるでいま悪い夢を見ているみたい
⑤ まるで幽霊に出くわしたかのよう
⑥ <u>私があなただったら</u>，そのやりがいのある仕事に立ち向かうだろう。
⑦ <u>もし，その装置がうまく作動しなかったら</u>，新しいものと交換すべきだ。
⑧ いま北風が吹いていたら，もっと速く進むことができるだろう。
⑨ <u>もし，あのとき天候が悪くなかったら</u>，そのような状況を避けることができただろう。

⑩ 若いときにもっと真剣に勉強していれば，そのときテストに合格できただろう。

リーディング
世界の商船隊はグローバルな職場で，多くの異なる国籍からなるクルーが共に航海をする長い伝統がある。世界の商船船舶の3分の2は多国籍かつ多言語のクルーを乗せている。それゆえに，コミュニケーションそして言語は多国籍企業がビジネスを適切に行い，グローバルな事業で役割を果たすために，極めて重要な要素となっている。コミュニケーション力および言語能力の欠如は日々の情報伝達を困難にし，これによって誤解が生じ，関係者すべてを巻き込む形で船の運航を危険にさらす可能性がある。

Story 13

練習A
① (What) (does) your brother (do)?
   My brother (works) (for) a major shipping company in Japan.
② (What) (does) your younger sister (do)?
   My sister (goes) to high school.
③ (What) (is) your (job/occupation)?
   I (am) an (engineer).
④ (Who) do you (work) (for)?
   I (work) for National Institute for Sea Training.
⑤ (Where) do you work?
   I (work) (in) Yokohama.

練習B
① I work (for) (a) (port) (radio) company.
② I (communicate) (with) (ships) [that come in and go out of Kosen port].
③ I feel (as) (if) I (worked) at an airport control tower.
④ I (ask) (the) (ship) what their estimated time of (arrival) or (departure) is.
⑤ I (relay) important information such as (weather) or (traffic) (condition) in the harbor (to) (the) (ship).

練習C
① First, I (stand) (watch).
② I am in (charge) of items such as (firefighting) (equipment), (lifeboats), and various other (emergency) items.
③ I am also (responsible) for (keeping) a (logbook).

④ I must work so as to (keep/maintain) a good working (relationship) among (officers), (engineers) and the crew.

⑤ Last but not least, I must learn (how) (to) (navigate) our ship.

練習D
① : ウ　② : ク　③ : イ　④ : ア　⑤ : カ　⑥ : ケ　⑦ : キ　⑧ : エ　⑨ : オ　⑩ : コ

練習E
Friend : (Long) (time) no (see).
　　　　So good to see you!

Mio 　 : Good to see you, too.
　　　　It (has) (been) three years (since) I saw you last.

Friend : (How) (have) (you) (been)?

Mio 　 : Very good. Since I came back from Hawaii, I have tried various things.
　　　　I (have) (worked) for a port radio company in Kosen port for a year.
　　　　How about you?

Friend : You (have) (had) good experiences in Japan.
　　　　I (have) also (experienced) various things.

リーディング

サンディエゴ統一港湾地区は，商船および湾内の水先案内人また近くの無線局との連絡のために，シェルター島にある港湾管制本部よりVHF-FM 無線局を運営している。チャンネル16(156.80MHz)は海難および緊急連絡や安全に関するメッセージ用と呼び出しのために使われる。チャンネル12(156.60MHz)は港湾業務のために使われる。無線局の呼び出し符号はKJC-824である。

Story 14

練習A
① あなたはすでに針路，速度そして天候に関する観測のすべてを正確に記入しています。

② 船長が遭難信号の発信の指示を出したところです。

③ おおよその損害のリストと，損害箇所の見取り図，そして調査報告の写しをここに同封してあります。

④ 修理の請求書はまだ我々の元には届いていません。

練習B
① 原油洗浄された積荷タンク

② 炭酸ガス消火器が使われたばかりの

③ 二酸化炭素が充満していた
④ すべての古いパッキンが取り除かれているということ
⑤ （運航に）必要なスラスタの性能が正確に把握されているならば
⑥ 稼働時間が約2万4000時間経過するまで

練習C
① 試してみてはどうですか？
② 冷却水タンクはどうですか？

練習D
① How can I (make them laugh or smile)?
② (How about showing them pictures) of your family or your friends?

練習E
① (Where in Japan are you from)?
② I (have never been to Toyama).
③ (How about going there) together when we get off the ship?

練習F
① explain　② pay　③ making　④ going　⑤ repeat　⑥ How about

リーディング

改正された1974年のSOLAS条約のV/28規則は，国際海事機関によって採択された提言を考慮し，国際運航に携るすべての船舶に，安全な航海上，重要でありその航海の完全な記録の復元のために必要な十分な詳細を含む運航業務および付随業務の記録をつけることを義務づけている。本決議は以下のような出来事を記録する上での指針を提供することを目的とする。

1) 航海に関連する情報の記録
国によって定められる要件だけでなく，必要に応じて，以下の業務や事項を記録することを推奨する。

1.1) 航海開始前
人員配置や食料などの積み込み，積荷，喫水，検査が行われた場合の船の安定性，ひずみの結果，制御機器の検査，操舵装置，航海計器，無線通信機器の点検など，船の状態に関係するすべてのデータの詳細が把握されそして記録されなくてはならない。

1.2) 航海中
針路，航海距離，位置決定，天候と海の状態，航海計画の変更，水先人の乗船／下船の詳細，航路指定方式または報告システムの対象でまたそれにのっとった水域への進入などの航海に関する詳細が記録されなくてはならない。

1.3) 特別な事象

船員または乗客の死亡や怪我，船上の装備や航行援助装置などの故障，潜在的に危険な状況，緊急事態や遭難信号の受信などの特別な事象もその詳細が記録されなくてはならない。

1.4) 船が停泊中または港にいる際

運航上および管理上の事柄や，船の安全や保安に関係する詳細が記録されなくてはならない。

Story 15

練習A

① つねに所持していなくてはならない多くの重要な書類があります。
② 正船尾，距離1.5マイルに，同航船があります。
③ 船上には35名おり，うち数人が重度の火傷を負っています。
④ 漏れや異音はあるか？ いいえ，ありません。
⑤ （これらが）入国・通関関連の書類です。
⑥ 船員名簿はありますか？ はい，これです，どうぞ。
⑦ パスポートはありますか？ はい，こちらです。

練習B

① Opening the strainer bypass valve, and closing the sea water inlet and outlet valves.
② Relieving the pressure inside the housing and dismantling the cover.
③ Changing to VHF channel 22.
④ Standing by on VHF channel 16.

練習C

① length　② width　③ depth　④ height

練習D

① The (length) of the ship is (53.59m).
② The (width) of the ship is (10.00m).
③ The (depth) of the ship is (5.40m).

練習E

① This ship is (241m long).
② This ship is (29.6m wide).

リーディング
　遭難通信は，遭難している人や航空機または船舶によって要請された医療援助を含む緊急の援助に関するすべての通信を含む。遭難通信には探索救難通信や現場通信も含まれることがある。遭難信号は他のどの通信よりも絶対的な優先権を持ち，遭難通信を受信したものはその信号を妨げる可能性のあるすべての通信を直ちに停止し，その遭難通信に使われた周波数に耳を傾けなければならない。

Story 16
練習A
① その書類はあなたが黄熱病に対するワクチンを接種したことを示すものです。
② 点呼名簿は，船橋，機関室，そして船員の居住区を含む船のいたるところに貼られている。
③ 対水速度は11.5ノットです。
④ With this shore pass, a crew will be given access through gates only for the area [that the vessel is berthed].
　ショアパスを持っていると，船が着岸しているエリアのゲートを通ることが許される。
⑤ Do you detect any temperature or pressure outside the normal parameters?
　正常値を超えた温度や圧力は見受けられるか？
⑥ Then, relieve the pressure inside the housing and dismantle the cover.
　そうしたら，ケーシング内の圧力を逃がし，それからカバーを取り外しなさい。

練習B
① 推測／イメージ：離れた状態を保つ
　単語の意味：控える，自制する，禁酒する
② 推測／イメージ：離れるように導く
　単語の意味：誘拐する，拉致する

練習C
① 不足，欠品　② 荷物，小包，輸送物　③ 使用法，使用量
④ 発達，発展　⑤ 議論　　　　　　　⑥ 船積み，発送，輸送

練習D
① 自動詞　② 自動詞　③ 他動詞　④ 他動詞　⑤ 他動詞
⑥ 他動詞　⑦ 他動詞　⑧ 他動詞　⑨ 他動詞　⑩ 他動詞

リーディング
機関当直の遂行
機関室が有人の場合，機関当直に責任ある士官は，方向や速度の変更の必要性に応じて，つねにただちに推進装置を操作できなくてはならない。機関室が定期的な無人運転状態にある場合，指名された機関当直士官は，機関室を見張るためにすぐに対応できる状態で待機していなくてはならない。

Story 17
練習A
① passport　② customs　③ immigration　④ shore pass　⑤ quarantine

練習B
① I boarded this ship in Spain.
② He will get off the ship at the next port.
③ The ship collided, and ten crew members on board the ship were injured.
④ The ship disembarked all crewmembers in Kosen port.

練習C
① Make sure that all crew walk on the designated walkways within the port boundary.
② Let's check the main engine next.
③ All crew must be familiar with emergency procedures.
④ We are ready to let go anchor.
⑤ Please relay this message to the captain.
⑥ Ensure a berth for the incoming ship.

リーディング
上陸する前に，税関，農務省および出入国管理局の通関手続きを終えなければなりません。

- 船上で待機してください。農務省または税関の検査官以外はあなたの船に乗船することは許可されていません。完全な許可を受けるまでいかなる人，動物，物品も船を離れることは許されていません。
- 到着時刻によって，その翌日に出入港手続きを完了するまで，税関および農務省はすべての船員の一晩船上待機を要求します。
- オーストラリア海域航海中または停泊中での廃棄物や食料品の海洋投棄はしてはいけません。指定されたバイオセキュリティ廃棄地点を利用してください。

- 農務省の検査官による検査が済むまで，すべての食料品および動物を外に持ち出さないでください。
- 他の船舶と食料品のやりとりをしないでください。
- 船舶を虫類から守ってください。

Story 18

練習A

We hereby report that based on the surveyor's recommendation, necessary repairs were ordered and completed, and that the ship was granted the seaworthy certificate for our homebound voyage.

We arrived here in Spain at 10:00 a.m. on 5th August and after discharging entire cargo, we underwent a survey by Mr. Park, of the Global Marine Surveyors Ltd. In accordance with his recommendations, we ordered necessary repairs by Pacific Dockyard Co. Ltd. Upon completion of the repairs on the 8th of August, Mr. Park granted the seaworthy certificate for our homebound voyage to Nagoya, Japan.

I have enclosed our damage list, sketch of damaged parts, and copies of each survey report.

The bill of repairs has not reached us yet, however, we will be in touch with you as soon as we receive it.

If you have any questions or concerns regarding this matter, please feel free to contact me.

練習B

① We have (completed) checking the main engine.
② I will (repair) the cooling water tank today.
③ I have (enclosed) a picture of my children with this letter.
④ Necessary parts and tools need to be (ordered) immediately.
⑤ The injured crewmember will (disembark) at the next port and (undergo) a surgery there.
⑥ The ship is going to (discharge) all the crew members at the next port.
⑦ We were (granted) a permission to access the controlled area of the harbor.

練習D

① (主語S) goods　(述語動詞V) are delivered
② (主語S) the strainer　(述語動詞V) is clogged
③ (主語S) no suspicious persons or unidentified vessels　(述語動詞V) are

④（主語S）all the fire detection sensors　（述語動詞V）are switched on
⑤（主語S）we　（述語動詞V）are
⑥（主語S）all crew　（述語動詞V）keep
⑦（主語S）the ship　（述語動詞V）was granted
⑧（主語S）you　（述語動詞V）must wear
⑨（主語S）safety rules　（述語動詞V）vary
⑩（主語S）generators　（述語動詞V）change

練習E

【We hereby report that based on the surveyor's recommendation, necessary repairs were ordered and completed, and that the ship was granted the seaworthy certificate for our homebound voyage.】

【We arrived here in Spain at 10:00 a.m. on 5th August and after discharging entire cargo, we underwent a survey by Mr. Park, of the Global Marine Surveyors Ltd. In accordance with his recommendations, we ordered necessary repairs by Pacific Dockyard Co. Ltd. Upon completion of the repairs on the 8th of August, Mr. Park granted the seaworthy certificate for our homebound voyage to Nagoya, Japan.】

【I have enclosed our damage list, sketch of damaged parts, and copies of each survey report.

The bill of repairs has not reached us yet, however, we will be in touch with you as soon as we receive it.

If you have any questions or concerns regarding this matter, please feel free to contact me.】

練習F

① We hereby report that the repairs you ordered have been completed.
② We hereby report that the ship has departed the Kosen port.
③ We hereby inform that our arrival will be one day delayed.
④ I am writing this e-mail to inquire about your new product.

練習H

① Please find the enclosed invoice.
② If you have any questions or concerns, please feel free to contact us.
③ Please let us know when the repairs will be completed.
④ Please see the attached report.

リーディング

5月20日午後8時30分にここに到着し、すべての貨物の陸揚げ完了後、ロイズ社の（耐航性）検査官エリオット氏の検査を受けました。また彼の勧告に従い、メッサーB&V社の造船所で必要な修理に出したことをここに報告申し上げます。

5月23日、エリオット氏の納得のいくよう修理が終了し、彼は母港に帰港する耐航性証明書を許可しました。

大まかな損害リスト、損害箇所の見取り図、それぞれの調査報告書の写しをここに（喜んで）同封します。

修理の請求書がまだこちらには届いていないのですが、受領後ただたに、そちらにお送りします。

Story 19

練習B
① We will berth (port) side (alongside).
② We will let go (starboard) anchor.
③ There is an (outbound) vessel in 10 o'clock direction.
④ Our (ETA) is 0800 hours (local) (time).

練習C
① one six　② zero five five　③ zero five seven　④ two five point zero
⑤ one five

練習D
① Fifteen plus thirty five equals fifty.
② Twelve added to eight equals twenty.
③ Eighty three minus twenty seven is fifty six.
④ Twenty seven substracted from eighty three is fifty six.
⑤ Thirty two times six equals one hundred ninety two.
⑥ Thirty two multiplied by six is one hundred ninety two.
⑦ Seventy two divided by eight is nine.

リーディング

国際海事機関（IMO）の標準海事通信用語集（SMCP）は、陸-船間（またその逆）、船舶間、そして船上での口頭でのコミュニケーションのなかでも最も重要な安全に関する分野を対象に作成された用語を含む。その目的は海上での言語の障壁を避け、事故を生みかねない誤解を回避することにある。

IMOのSMCPは英語の基礎的な知識に基づいていて、海事英語の簡易版として作成

された。用語集には，緊急的な状況に使われる決まり文句や応答だけでなく係留などの日常的な状況が含まれる。

Story 20

練習A
- ① Do you know who is on watch now?
- ② Please explain why they are late.
- ③ Could you please show me how you clean the strainer?
- ④ I am not sure if this is yours.
- ⑤ The captain asked if I understood.
- ⑥ They don't know what yose-nabe is.
- ⑦ I ask the ship what their ETA is.

練習B
- ① ⑦ No problem.
- ② ④ It was my honor to have you as my student.
- ③ ⑦ Anytime!

練習E

解答例。下線部は言い換え可能な語句。

(トピック) Thank you so much for everything. I learned a lot from you.

(具体例) For example, you taught me how to keep watch <u>in the engine room</u> (<u>on the deck</u>). And, I learned a procedure of how to <u>detect abnormalities</u> (<u>clean the strainer</u> / <u>prepare various equipment for safety and emergency</u>). The most important thing is to communicate with all crew on board. And all of you were very kind to me and helped me so often.

(まとめ) I really appreciate your kindness.

リーディング

ガイドラインB2.4.4 — 年次休暇の取得

1. 年次休暇を取る時期は，条例，団体協約，仲裁裁定もしくは国の慣行と矛盾しないその他の方法により定められていない限り，関係する船員もしくはその代表者との協議の上で，できるだけ彼らの意向に一致する形で船主によって決定されなければならない。
2. 船員は原則として当人に実質上つながりのある場所，つまり通常帰還する資格のある場所と同じ場所で，年次休暇を取る権利を有する。船員は，雇用契約または国の法律あるいは条例で定められている場合を除いて，当人の同意なしに，

別の場所で、取得すべき年次休暇を取るように命じられるべきではない。
3. もし、船員がこのガイドラインの第2項で定められている場所以外で年次休暇を取らなくてはならない場合、その船員は契約している場所または雇用された場所のどちらか本国により近い場所へ、無料で移動する資格が与えられるべきである。移動に直接含まれる費用やその他の費用は船主の負担とすべきであり、移動時間は船員の取得すべき有給年次休暇から差し引かれてはならない。
4. 年次休暇を取っている船員は、極度の緊急事態の場合のみ、船員の同意を得て、呼び戻されるものとする。

【編著者の紹介】

池田 恭子（いけだ きょうこ）
通訳、翻訳者、英語教師として仕事をしてきた経験をもとに学校や学習者のニーズに合わせた海外での研修プログラムの企画などを行う。平成24年4月からは、ハワイ大学カウアイコミュニティカレッジで国際教育プログラムを担当する。

長山 昌子（ながやま あきこ）
平成8年8月1日に富山商船高等専門学校（現・富山高等専門学校）に赴任。教養科に所属し、英文法や英語会話を担当し、学生指導にあたる。平成24年4月からは、大学間連携共同教育推進事業「海事分野における高専・産業界連携による人材育成システムの開発」の特命教授として、教材開発に携わり、学生の指導にあたる。

ISBN978-4-303-23346-4

マリタイムカレッジシリーズ
Navigating English

2015年3月30日　初版発行　　　　　　　　　　　　ⓒK. IKEDA／A. NAGAYAMA　2015

編著者　池田恭子・長山昌子　　　　　　　　　　　　　　　　　　　　検印省略
発行者　岡田節夫
発行所　海文堂出版株式会社
　　　　本社　東京都文京区水道2-5-4（〒112-0005）
　　　　　　　電話 03（3815）3291代　FAX 03（3815）3953
　　　　　　　http://www.kaibundo.jp/
　　　　支社　神戸市中央区元町通3-5-10（〒650-0022）
日本書籍出版協会会員・工学書協会会員・自然科学書協会会員

PRINTED IN JAPAN　　　　　　　　　　　　印刷　田口整版／製本　誠製本

JCOPY　＜(社)出版者著作権管理機構　委託出版物＞
本書の無断複写は著作権法上での例外を除き禁じられています。複写される場合は、そのつど事前に、(社)出版者著作権管理機構（電話 03-3513-6969, FAX 03-3513-6979, e-mail: info@jcopy.or.jp）の許諾を得てください。